VOLUME ONE HUNDRED AND FORTY SIX

INTERNATIONAL REVIEW OF NEUROBIOLOGY

New Therapeutic Strategies for Brain Edema and Cell Injury

INTERNATIONAL REVIEW OF NEUROBIOLOGY

VOLUME 146

SERIES EDITOR

PATRICIA JANAK

Janak Lab, Dunning Hall
Baltimore, MD, USA

PETER JENNER

Division of Pharmacology and Therapeutics
GKT School of Biomedical Sciences
King's College, London, UK

VOLUME ONE HUNDRED AND FORTY SIX

INTERNATIONAL REVIEW OF NEUROBIOLOGY

New Therapeutic Strategies for Brain Edema and Cell Injury

Edited by

HARI SHANKER SHARMA
International Experimental Central Nervous System Injury & Repair (IECNSIR), Department of Surgical Sciences, Anesthesiology & Intensive Care Medicine, Uppsala University Hospital, Uppsala University, Uppsala, Sweden

ARUNA SHARMA
International Experimental Central Nervous System Injury & Repair (IECNSIR), Department of Surgical Sciences, Anesthesiology & Intensive Care Medicine, Uppsala University Hospital, Uppsala University, Uppsala, Sweden

ELSEVIER

ACADEMIC PRESS
An imprint of Elsevier

Academic Press is an imprint of Elsevier
50 Hampshire Street, 5th Floor, Cambridge, MA 02139, United States
525 B Street, Suite 1650, San Diego, CA 92101, United States
The Boulevard, Langford Lane, Kidlington, Oxford OX5 1GB, United Kingdom
125 London Wall, London, EC2Y 5AS, United Kingdom

First edition 2019

ISBN: 978-0-12-816754-0
ISSN: 0074-7742

Publisher: Zoe Kruze
Acquisition Editor: Sam Mahfoudh
Editorial Project Manager: Peter Llewellyn
Production Project Manager: Abdulla Sait
Cover Designer: Greg Harris

Typeset by SPi Global, India

Contents

Contributors

Mario Antonazzo
Department of Pharmacology, Faculty of Medicine and Nursing, University of the Basque Country (UPV/EHU), Leioa; Neurodegenerative Diseases Group, BioCruces Bizkaia Health Research Institute, Barakaldo, Bizkaia, Spain

Asier Aristieta
Université de Bordeaux; Centre National de la Recherche Scientifique, Institut des Maladies Neurodégénératives, UMR 5293, Bordeaux, France

Harkaitz Bengoetxea
Department of Neurosciences, Faculty of Medicine and Nursing, University of the Basque Country (UPV/EHU), Leioa, Spain

María Botta
Department of Pharmacology, Faculty of Pharmacy, University of the Basque Country (UPV/EHU), Vitoria-Gasteiz, Spain

Prasanta Kumar Dey
Neurophysiology Research Unit, Department of Physiology, Institute of Medical Sciences, Banaras Hindu University, Varanasi, Uttar Pradesh, India

Lianyuan Feng
Department of Neurology, Bethune International Peace Hospital, Shijiazhuang, China

T. Fernández-Valle
Movement Disorders and Autonomic Unit, Neurology Department, Cruces University Hospital; Neurodegenerative Unit, Biocruces Health Research Institute, Barakaldo, Spain

I. Gabilondo
Movement Disorders and Autonomic Unit, Neurology Department, Cruces University Hospital; Neurodegenerative Unit, Biocruces Health Research Institute, Barakaldo, Spain

J.C. Gómez-Esteban
Movement Disorders and Autonomic Unit, Neurology Department, Cruces University Hospital; Neurodegenerative Unit, Biocruces Health Research Institute, Barakaldo, Spain

Eugene A. Kiyatkin
Behavioral Neuroscience Branch, National Institute on Drug Abuse—Intramural Research Program, National Institutes of Health, DHHS, Baltimore, MD, United States

Jose V. Lafuente
Department of Neurosciences, Faculty of Medicine and Nursing, University of the Basque Country (UPV/EHU), Leioa; Nanosurgery, Biocruces Health Research Institute, Barakaldo, Spain

Cristina Miguelez
Department of Pharmacology, Faculty of Medicine and Nursing, University of the Basque Country (UPV/EHU), Leioa; Neurodegenerative Diseases, Biocruces Health Research Institute, Barakaldo, Spain

Teresa Morera-Herreras
Department of Pharmacology, Faculty of Medicine and Nursing, University of the Basque Country (UPV/EHU), Leioa; Neurodegenerative Diseases Group, BioCruces Bizkaia Health Research Institute, Barakaldo, Bizkaia, Spain

Dafin Fior Muresanu
Department of Clinical Neurosciences, University of Medicine & Pharmacy; "RoNeuro" Institute for Neurological Research and Diagnostic, Cluj-Napoca, Romania

Feng Niu
CSPC NBP Pharmaceutical Medicine, Shijiazhuang, China

Ala Nozari
Anesthesia and Critical Care; Anesthesiology & Intensive Care, Massachusetts General Hospital, Boston, MA, United States

Elena Paredes-Rodriguez
Department of Pharmacology, Faculty of Medicine and Nursing, University of the Basque Country (UPV/EHU), Leioa; Neurodegenerative Diseases, Biocruces Health Research Institute, Barakaldo, Spain

José Ángel Ruiz-Ortega
Department of Pharmacology, Faculty of Medicine and Nursing, University of the Basque Country (UPV/EHU), Leioa; Department of Pharmacology, Faculty of Pharmacy, University of the Basque Country (UPV/EHU), Vitoria-Gasteiz; Neurodegenerative Diseases Group, BioCruces Bizkaia Health Research Institute, Barakaldo, Bizkaia, Spain

Seaab Sahib
Cell & Molecular Biology Program (CEMB), Department of Chemistry and Biochemistry, University of Arkansas—Fayetteville, Fayetteville, AR, United States

Aruna Sharma
International Experimental Central Nervous System Injury & Repair (IECNSIR), Department of Surgical Sciences, Anesthesiology & Intensive Care Medicine, Uppsala University Hospital, Uppsala University, Uppsala, Sweden

Hari Shanker Sharma
International Experimental Central Nervous System Injury & Repair (IECNSIR), Department of Surgical Sciences, Anesthesiology & Intensive Care Medicine, Uppsala University Hospital, Uppsala University, Uppsala, Sweden

Mark A. Smith
Department of Pathology, Case Western Reserve University, Cleveland, OH, United States

Z. Ryan Tian
Cell & Molecular Biology Program (CEMB), Department of Chemistry and Biochemistry, University of Arkansas—Fayetteville, Fayetteville, AR, United States

Luisa Ugedo
Department of Pharmacology, Faculty of Medicine and Nursing, University of the Basque Country (UPV/EHU), Leioa; Neurodegenerative Diseases, Biocruces Health Research Institute, Barakaldo, Spain

Sergio Vegas-Suarez
Department of Pharmacology, Faculty of Medicine and Nursing, University of the Basque Country (UPV/EHU), Leioa; Neurodegenerative Diseases, Biocruces Health Research Institute, Barakaldo, Spain

Lars Wiklund
International Experimental Central Nervous System Injury & Repair (IECNSIR), Department of Surgical Sciences, Anesthesiology & Intensive Care Medicine, Uppsala University Hospital, Uppsala University, Uppsala, Sweden

Preface

Brain edema also known as cerebral edema is one of the fundamental outcomes of all kinds of noxious insults to the central nervous system (CNS) leading to serious clinical consequences (Klatzo, 1967; Plum, 1966; Stern, 1965). Excessive edematous swelling within the closed cranial vault could compress the vital centers of the brain resulting in instant death (Mendelow & Teasdale, 1983; Schmidt, Heyde, Ertel, & Stahel, 2005).

Brain edema is defined by excessive accumulation of fluid within the extracellular or intracellular compartments of the brain causing alterations in the cellular structures inflicting morbidity and mortality (Reulen, 1976; Reulen, Medzihradsky, Enzenbach, Marguth, & Brendel, 1969).

About 250 years ago, Vesalius described subtle changes in brain parenchyma caused by cerebral ventricular dilation due to excess accumulation of fluid in the brain (Vesalius, 1543). This observation led to Robert Whytt to formulate the concept of hydrocephalus caused by excess accumulation of intracranial water leading to enlarge cerebral ventricles (Whytt, 1768).

This led to the concept of "cerebral edema" as mentioned by Cheyne confirming Whytt these hydrocephalus cases contained excessive water in the brain itself and the fluid was driven from an abnormal cerebral vasculature (Cheyne, 1814; Lyons, 1995). This was followed by more research on brain circulation notably by Alexander Monro who described Foramen of Monro linking lateral and third ventricle in the brain and put forward his theory of intracranial blood volume in 1783 (Monro, 1783). This idea was only substantiated by Kellie (1824a) known as the famous Monro-Kellie doctrine. The doctrine states that the CNS and its accompanying fluids are enclosed in a rigid container whose total volume, i.e., brain, blood, or cerebrospinal fluid, tends to remain constant. Any increase in the volume of one component will elevate pressure and decrease the volume of one of the other elements (Kellie, 1824b). This principle was supported by Abercrombie in 1828 (Abercrombie, 1828; No authors listed, 1828). However, Abercrombie did not believe that substantial effusion of fluid might come from blood to the brain because he did not find a reduction in the blood volume in cases of cerebral edema. Abercrombie believed that the softness of brain tissue is due to degeneration and not because of fluid accumulation (Torack, 1982).

Thus, the concept of cerebral edema as a medical entity disappeared from medicine books until 1898 when Mills published a book mentioning Cerebral Edema in 3 pages citing a report by Preston (1893) in which 68 cases of cerebral edema were described (Mills, 1898; Preston, 1893). Mills and Preston using microscopy described the concept of cerebral edema stating that if the brain is infiltrated with serum, then it is usually pale and soggy—the wet brain (Mills, 1898; Preston, 1893). Microscopic images could see dilatation and distention of the perivascular space. Although this observation sets the modern concept of brain edema, this work was completely ignored until next century. In 1919, Reichardt who developed methods to measure normal brain volume using brain–cranium ratio showed that any reduction of the normal 8% difference indicates brain swelling around a focal lesion, brain tumor, or abscess (Reichardt, 1919). In 1929, Spatz differentiated between swelling and edema considering swelling as being enlarged, firm, and dry, whereas edema was enlarged, soft, and wet (Spatz, 1929). Spatz also stated that both results due to fluid problem and reestablished cerebral edema as a pathological entity (Spatz, 1929).

In 1965, Lee and Bakay (Bakay, 1965; Lee & Bakay, 1965) used an electron microscope to study cerebral edema followed by Reese and Karnovsky (1967) and demonstrated that the morphological sites of the blood–brain barrier (BBB) for protein are the tight junctions between the endothelial cells that are formed due to fusion of adjacent endothelial cell linings. These findings were further confirmed by Brightmann et al. at the ultrastructural level (Brightman, Klatzo, Olsson, & Reese, 1970). Alterations in the BBB function result in extravasation of serum proteins into the brain fluid microenvironment causing edema formation. This forms the concept of "vasogenic brain edema" a separate entity from "cytotoxic edema" by Igor Klatzo and his team (Klatzo, Piraux, & Laskowski, 1958; Steinwall & Klatzo, 1965).

However, the differentiation of these two fundamentally different forms of brain edema vasogenic and cytotoxic defined by Klatzo (1967) could exist together in many cases. In vasogenic edema leakage of plasma proteins and vascular damage due to breakdown of the BBB are prominent (Brightman et al., 1970; Klatzo et al., 1958; Steinwall & Klatzo, 1965). However, extravasation of water and ions is primarily responsible for the development of cytotoxic edema (Klatzo, 1967; Steinwall & Klatzo, 1965). These two different types of brain edema either coexist in some cases or one form of edema could lead to the other form of edema in several conditions of brain pathology (Reulen, Graham, Spatz, & Klatzo, 1977). However, vasogenic edema is the most common form of cerebral edema seen in a wide variety of pathological insults to the CNS (Klatzo et al., 1958; Reulen, 1976).

Most of our knowledge about the progression, persistence, and resolution of brain edema comes from experimental observations caused by cold lesion where a cold metal rod (−70 to 80 °C) was used for cortical injuries in several mammalian species (Bothe, Bodsch, & Hossmann, 1984; Reulen et al., 1977; Smith & Marque, 1976). Additional knowledge of cerebral edema comes from electron microscopic studies of peritumoral edema and traumatic injuries (Bartkowski, 1984; Black, Hoff, McGillicuddy, & Gebarski, 1986; Hardman, 1979; Waga, Tochio, & Sakakura, 1979). These studies show the enlargement of extracellular space, and an increase in local interstitial fluid pressure together with tissue pressure gradients plays a dominant role in the spread of edema across the CNS (Reulen et al., 1977). The tissue pressure gradient allows a greater increase in edema spread in white matter because of a loosely arranged texture of nerve fibers than the gray matter. The gray matter is an interwoven network of complex cellular structures and nerve fibers that is more resistant in allowing spread of edema in the CNS (Reulen, 1976; Reulen et al., 1977, 1969).

Further studies on cerebral edema showed that ischemic infarcts and intracerebral hemorrhages could result in more than 80% mortality due to brain swelling. Other causes of brain edema with lifetime disability may occur in stroke, hypertensive encephalopathy, and acute liver failure. Traumatic brain injury is another important factor with mortality greater than 50% and is also associated with severe brain edema (see Stokum, Gerzanich, & Simard, 2016).

Brain edema research is carried out in our laboratory since the 1980s in several experimental animal models. In arterial hypertension cases massive vasogenic edema is caused by intracarotid infusion of prostaglandin synthesis inhibitor that was associated with BBB breakdown due to acute loss of cerebral autoregulation (Dey, Sharma, & Rao, 1980). In 1983 and 1984, our studies show that brain edema associated with traumatic brain injury and closed head injury is influenced by external hot or cold environment (Dey & Sharma, 1983, 1984). Traumatic brain edema is also associated with neurotransmitters like serotonin and prostaglandin that participate in the BBB breakdown and induce vasogenic edema (Dey & Sharma, 1983, 1984).

In 1990, our studies are the first to show that breakdown of the BBB is also associated with hyperthermic brain injury, not described earlier (Sharma & Cervós-Navarro, 1990). Later studies showed that hyperthermic brain injury-induced edema showing both cytotoxic and vasogenic types in the same brain areas (Sharma, Westman, & Nyberg, 1998). Discoveries of hyperthermia-induced brain edema prompted us to evaluate drugs of abuse that normally induce brain hyperthermia causing BBB breakdown and brain edema formation. This hypothesis was indeed found true after acute administration of methamphetamine, cocaine, and 3,4-methylenedioxymethamphetamine (MDMA) as well as following morphine withdrawal in rats and mice (Sharma & Ali, 2006, 2008; Sharma & Kiyatkin, 2009; Sharma, Muresanu, Sharma, & Patnaik, 2009). These observations opened new vistas in brain edema research, and further showed that drugs of abuse-induced brain edema are also associated with environmental temperature (Sharma & Kiyatkin, 2009). Our recent research showed that nanoparticles from either silica dust or engineered nanoparticles from the metals also induced BBB breakdown and vasogenic edema formation together with behavioral disturbances (Sharma, Ali, Hussain, Schlager, & Sharma, 2009; Sharma, Patnaik, Sharma, Sjöquist, & Lafuente, 2009; Sharma & Sharma, 2007). Thus, efforts are needed to contain brain edema development during diverse kinds of CNS insults to enhance the quality of life of the victims using novel therapeutic strategies.

This volume is the first of its kind where these new developments and recent therapeutic strategies in treating brain edema development following various kinds of noxious insults to the CNS are presented in 10 selected peer-reviewed invited contributions by world-renowned experts in the field.

In Chapter 1, Aruna Sharma (Uppsala, Sweden) described 5-hydroxytryptophan (5-HTP)-induced brain edema and BBB breakdown and their modification with selected drugs. Since 5-HTP is used to treat depression, this is important to understand the pathophysiology of this compound in relation to brain edema.

Anesthetics may influence the disease process by itself and lead to cellular changes in the brain especially in developing brain. Whether anesthetic influence pathophysiology of traumatic brain injury in the adult brain is still unclear. This aspect is discussed by Hari Shanker Sharma (Uppsala University) in Chapter 2 in great details.

In Chapter 3 Seaab Sahib (Fayetteville, AR, USA) presented a new approach of titanium plate associated with TiO_2 nanowires implant that could help reducing spinal cord injury-induced edema and cell injury by enhancing spinal cord conduction when nanowired delivery of DL-3-*n*-butylphthalide was made topically over the injured spinal cord.

Eugene A. Kiyatkin (Baltimore, MD, USA) discussed methamphetamine-induced vasogenic brain edema that is aggravated at high ambient temperature and may worsen brain pathology in Chapter 4.

Teresa Morera-Herreras (Bilbao, Spain) in Chapter 5 discussed cannabinoids as neuroprotective agents in movement disorders related to edema and cell changes in the brain.

The role of serotoninergic system is discussed in Parkinson's disease induced brain pathology by Luisa Ugedo (Bilbao, Spain) in Chapter 6.

Finally in Chapter 7, T. Fernández-Valle presented new therapeutic targets to α-synuclein in treating brain pathology in Parkinson's disease.

We sincerely hope that these new avenues of brain edema therapy discussed in this volume will expand our knowledge and encourage new research in the field. The volume is thus indispensable for students, researchers, policy makers, neuroscience course administrators, teachers, and professionals alike. The new ideas emanating from this volume will further stimulate innovation and drug development for the benefit of mankind resulting in better therapeutic strategies in CNS edema and prolong healthy life!

Hari Shanker Sharma
Aruna Sharma
Uppsala University, Uppsala, Sweden

References

Abercrombie, J. (1828). Pathological and practical researches on the diseases of the brain and the spinal chord. *Edinburgh Medical and Surgical Journal, 29*(95), 329–357.

Bakay, L. (1965). Morphological and chemical studies in cerebral edema: Triethyl tin-induced edema. *Journal of the Neurological Sciences, 2*(1), 52–67 [No abstract available].

Bartkowski, H. M. (1984). Peritumoral edema. *Progress in Experimental Tumor Research, 27*, 179–190 [Review. No abstract available].

Black, K. L., Hoff, J. T., McGillicuddy, J. E., & Gebarski, S. S. (1986). Increased leukotriene C4 and vasogenic edema surrounding brain tumors in humans. *Annals of Neurology, 19*(6), 592–595.

Bothe, H. W., Bodsch, W., & Hossmann, K. A. (1984). Relationship between specific gravity, water content, and serum protein extravasation in various types of vasogenic brain edema. *Acta Neuropathologica, 64*(1), 37–42.

Brightman, M. W., Klatzo, I., Olsson, Y., & Reese, T. S. (1970). The blood-brain barrier to proteins under normal and pathological conditions. *Journal of the Neurological Sciences, 10*(3), 215–239 [Review. No abstract available].

Cheyne, J. (1814). *An essay on hydrocephalus acutus*. Philadelphia: Finley.

Dey, P. K., & Sharma, H. S. (1983). Ambient temperature and development of traumatic brain oedema in anaesthetized animals. *The Indian Journal of Medical Research, 77*, 554–563 [No abstract available].

Dey, P. K., & Sharma, H. S. (1984). Influence of ambient temperature and drug treatments on brain oedema induced by impact injury on skull in rats. *Indian Journal of Physiology and Pharmacology, 28*(3), 177–186.

Dey, P. K., Sharma, H. S., & Rao, K. S. (1980). Effect of indomethacin (a prostaglandin synthetase inhibitor) on the permeability of blood-brain and blood-CSF barriers in rat. *Indian Journal of Physiology and Pharmacology, 24*(1), 25–36.

Hardman, J. M. (1979). The pathology of traumatic brain injuries. *Advances in Neurology, 22*, 15–50 [Review. No abstract available].

Kellie, G. (1824a). Reflections on the pathology of the brain: Part II. *Transactions of the Medico-Chirurgical Society of Edinburgh, 1*, 123–169 [No abstract available].

Kellie, G. (1824b). An account of the appearances observed in the dissection of two of three individuals presumed to have perished in the storm of the 3d, and whose bodies were discovered in the vicinity of Leith on the morning of the 4th November 1821; with some reflections on the pathology of the brain: Part I. *Transactions of the Medico-Chirurgical Society of Edinburgh, 1*, 84–122.

Klatzo, I. (1967). Presidental address. Neuropathological aspects of brain edema. *Journal of Neuropathology and Experimental Neurology, 26*(1), 1–14 [Review. No abstract available].

Klatzo, I., Piraux, A., & Laskowski, E. J. (1958). The relationship between edema, blood-brain-barrier and tissue elements in a local brain injury. *Journal of Neuropathology and Experimental Neurology, 17*(4), 548–564 [No abstract available].

Lee, J. C., & Bakay, L. (1965). Ultrastructural changes in the edematous central nervous system. 1. Triethyltin edema. *Archives of Neurology, 13*, 48–57 [No abstract available].

Lyons, J. B. (1995). John Cheyne's classic monographs. *Journal of the History of the Neurosciences, 4*(1), 27–35.

Mendelow, A. D., & Teasdale, G. M. (1983). Pathophysiology of head injuries. *The British Journal of Surgery, 70*(11), 641–650 [Review. No abstract available].

Mills, C. K. (1898). *The nervous system and its diseases* (pp. 438–441). Philadelphia: JB Lippincott Co.

Monro, S. A. (1783). *Observations on the structure and functions of the nervous system.* Edinburgh: W Creech.

No authors listed. (1828). Dr. Abercrombie on the brain and spinal cord. *The Medico-Chirurgical Review, 8*(16), 337–351 [Review. No abstract available].

Plum, F. (1966). Brain swelling and edema in cerebral vascular disease. *Research Publications—Association for Research in Nervous and Mental Disease, 41*, 318–348 [Review. No abstract available].

Preston, G. J. (1893). Cerebral edema. *The Journal of Nervous and Mental Disease, 21*, 494–503.

Reese, T. S., & Karnovsky, M. J. (1967). Fine structural localization of a blood-brain barrier to exogenous peroxidase. *The Journal of Cell Biology, 34*(1), 207–217.

Reichardt, M. (1919). Hirnschwellung. *Allgemeine Zeitschrift fur Psychiatrie, 75*, 34–103.

Reulen, H. J. (1976). Vasogenic brain oedema. New aspects in its formation, resolution and therapy. *British Journal of Anaesthesia, 48*(8), 741–752 [Review. No abstract available].

Reulen, H. J., Graham, R., Spatz, M., & Klatzo, I. (1977). Role of pressure gradients and bulk flow in dynamics of vasogenic brain edema. *Journal of Neurosurgery, 46*(1), 24–35.

Reulen, H. J., Medzihradsky, F., Enzenbach, R., Marguth, F., & Brendel, W. (1969). Electrolytes, fluids, and energy metabolism in human cerebral edema. *Archives of Neurology, 21*(5), 517–525 [No abstract available].

Schmidt, O. I., Heyde, C. E., Ertel, W., & Stahel, P. F. (2005). Closed head injury—An inflammatory disease? *Brain Research. Brain Research Reviews, 48*(2), 388–399 [Epub 2005 Jan 28. Review].

Sharma, H. S., & Ali, S. F. (2006). Alterations in blood-brain barrier function by morphine and methamphetamine. *Annals of the New York Academy of Sciences, 1074*, 198–224.

Sharma, H. S., & Ali, S. F. (2008). Acute administration of 3,4-methylenedioxymethamphetamine induces profound hyperthermia, blood-brain barrier disruption, brain edema formation, and cell injury. *Annals of the New York Academy of Sciences, 1139*, 242–258.

Sharma, H. S., Ali, S. F., Hussain, S. M., Schlager, J. J., & Sharma, A. (2009). Influence of engineered nanoparticles from metals on the blood-brain barrier permeability, cerebral blood flow, brain edema and neurotoxicity. An experimental study in the rat and mice using biochemical and morphological approaches. *Journal of Nanoscience and Nanotechnology, 9*(8), 5055–5072.

Sharma, H. S., & Cervós-Navarro, J. (1990). Brain oedema and cellular changes induced by acute heat stress in young rats. *Acta Neurochirurgica. Supplementum (Wien), 51*, 383–386.

Sharma, H. S., & Kiyatkin, E. A. (2009). Rapid morphological brain abnormalities during acute methamphetamine intoxication in the rat: An experimental study using light and electron microscopy. *Journal of Chemical Neuroanatomy, 37*(1), 18–32 [Epub 2008 Aug 19].

Sharma, H. S., Muresanu, D., Sharma, A., & Patnaik, R. (2009). Cocaine-induced breakdown of the blood-brain barrier and neurotoxicity. *International Review of Neurobiology, 88*, 297–334 [Review].

Sharma, H. S., Patnaik, R., Sharma, A., Sjöquist, P. O., & Lafuente, J. V. (2009). Silicon dioxide nanoparticles (SiO_2, 40-50 nm) exacerbate pathophysiology of traumatic spinal cord injury and deteriorate functional outcome in the rat. An experimental study using pharmacological and morphological approaches. *Journal of Nanoscience and Nanotechnology, 9*(8), 4970–4980.

Sharma, H. S., & Sharma, A. (2007). Nanoparticles aggravate heat stress induced cognitive deficits, blood-brain barrier disruption, edema formation and brain pathology. *Progress in Brain Research, 162*, 245–273 [Review].

Sharma, H. S., Westman, J., & Nyberg, F. (1998). Pathophysiology of brain edema and cell changes following hyperthermic brain injury. *Progress in Brain Research, 115*, 351–412 [Review. No abstract available].

Smith, A. L., & Marque, J. J. (1976). Anesthetics and cerebral edema. *Anesthesiology, 45*(1), 64–72.

Spatz, H. (1929). Die bedeutung de "Symptomatischen hirnschwellung" für de hirntumoren und für andere raumbeengende prozesse in der schdelgrube. *Archives of Psychiatry, 88*, 790–794.

Steinwall, O., & Klatzo, I. (1965). Double tracer methods in studies on blood-brain barrier dysfunction and brain edema. *Acta Neurologica Scandinavica. Supplementum, 13*(Pt 2), 591–595 [No abstract available].

Stern, W. E. (1965). The contribution of the laboratory to an understanding of the cerebral edemas. A review of recent progress. *Neurology, 15*(10), 902–912 [Review. No abstract available].

Stokum, J. A., Gerzanich, V., & Simard, J. M. (2016). Molecular pathophysiology of cerebral edema. *Journal of Cerebral Blood Flow and Metabolism, 36*(3), 513–538. https://doi.org/10.1177/0271678X15617172 [Epub 2015 Nov 16].

Torack, R. M. (1982). Historical aspects of normal and abnormal brain fluids. III. Cerebral edema. *Archives of Neurology, 39*(6), 355–357.

Vesalius, A. (1543). *De humani corporis labrica librorum epitome*. Baselieae: Joannis Opporini.

Waga, S., Tochio, H., & Sakakura, M. (1979). Traumatic cerebral swelling developing within 30 minutes after injury. *Surgical Neurology, 11*(3), 191–193.

Whytt, R. (1768). *Observations on the dropsy in the brain*. Edinburgh J Balfour.

Acknowledgments

We are grateful to Peter Jenner (London, UK) for the idea, encouragement and development of the book project in current form. We are deeply indebted to Sam Mahfoudh, Peter Llewellyn and Abdulla Sait from Elsevier for their constant inspiration, guidance, and support throughout during compilation of this volume. Without their active involvement and help, this volume in this form is not possible. We deeply appreciate their professional support and help. For computer assistance and graphic support, we thank Suraj Sharma, Blekinge Institute of Technology, Karlskrona, Sweden.

Hari Shanker Sharma
Aruna Sharma

CHAPTER ONE

5-Hydroxytryptophan: A precursor of serotonin influences regional blood-brain barrier breakdown, cerebral blood flow, brain edema formation, and neuropathology

Aruna Sharma[a], Mark A. Smith[b], Dafin Fior Muresanu[c,d], Prasanta Kumar Dey[e], Hari Shanker Sharma[a,*]

[a]International Experimental Central Nervous System Injury & Repair (IECNSIR), Department of Surgical Sciences, Anesthesiology & Intensive Care Medicine, Uppsala University Hospital, Uppsala University, Uppsala, Sweden
[b]Department of Pathology, Case Western Reserve University, Cleveland, OH, United States
[c]Department of Clinical Neurosciences, University of Medicine & Pharmacy, Cluj-Napoca, Romania
[d]"RoNeuro" Institute for Neurological Research and Diagnostic, Cluj-Napoca, Romania
[e]Neurophysiology Research Unit, Department of Physiology, Institute of Medical Sciences, Banaras Hindu University, Varanasi, Uttar Pradesh, India
*Corresponding author: e-mail addresses: sharma@surgsci.uu.se; hssharma@aol.com; harishanker_sharma55@icloud.com

Contents

International Review of Neurobiology, Volume 146
ISSN 0074-7742
https://doi.org/10.1016/bs.irn.2019.06.005

Abstract

5-Hydroxytryptophan (5-HTP), a precursor of serotonin, is therapeutically used for several psychiatric disorders such as anxiety and depression in the clinic. However, severe side effects, including abnormal mental functions, behavioral disturbances and intolerance are associated with this treatment. 5-HTP-induced elevation of plasma and brain serotonin levels may affect blood-brain barrier (BBB) breakdown, edema formation and regional cerebral blood flow (CBF) disturbances. Breakdown of BBB to serum proteins leads to vasogenic brain edema formation and cellular injuries. However, 5-HTP-neurotoxicity is still not well known. In this investigations 5-HTP induced elevation of endogenous plasma and brain serotonin levels and its effect on BBB breakdown, edema formation neuronal injuries was examined in a rat model. Furthermore, potential role of oxidative stress and nitric oxide (NO) was evaluated. In addition, several neurochemical agents such as *p*-CPA (5-HT synthesis inhibitor) indomethacin (prostaglandin synthase inhibitor), diazepam (ant stress drug), cyproheptadine, ketanserin (5-HT2 receptor antagonists) and vinblastine (inhibitor of microtubule function) were examined on 5-HT neurotoxicity. Our observations suggest that 4h after 5-HTP administrations, the endogenous serotonin levels increased by fourfold (150mg/kg) in the plasma and brain associated with profound hyperthermia (+3.86±0.24°C, oxidative stress and NO upregulation. Breakdown of the BBB to Evans blue albumin (EBA) in 8 brain regions and to $^{[131]}$Iodine in 14 brain regions was observed. The CBF exhibited marked reduction in all the brain regions examined. Brain edema and cellular injuries are present in the areas associated with BBB disruption. Drug treatments reduced the BBB breakdown, edema formation NO production and brain pathology. These observations are the first to point out that 5-HTP-neurotoxicity caused by BBB breakdown, edema formation and NO production is instrumental in causing adverse mental and behavioral abnormalities, not reported earlier.

1. Introduction

The blood-brain barrier (BBB) maintains the composition of extracellular fluid environment of the central nervous system (CNS) by strictly regulating the blood-brain interface in physiological conditions (Bradbury, 1979;

Rapoport, 1976; Sharma, 1982, 1999, 2009; Sharma & Westman, 2003, 2004). A slight alteration in the BBB function to blood-borne agents could adversely impair neuronal functions leading to long-term mental abnormalities (Bosma, van Noorden, Schlingemann, & Klaassen, 2018; Brightman, Klatzo, Olsson, & Reese, 1970; Erickson & Banks, 2018; Sharif et al., 2018; Sharma et al., 2016). This is further supported by the fact that in several neurological diseases and psychiatric disorders breakdown of the BBB and blood-cerebrospinal fluid barrier (BCSFB) is observed (Nation et al., 2019; Sweeney, Kisler, Montagne, Toga, & Zlokovic, 2018; Sweeney, Sagare, & Zlokovic, 2018; Sweeney, Zhao, Montagne, Nelson, & Zlokovic, 2019). Thus, dementia, depression, anxiety disorders, schizophrenia, autism as well as neurodegenerative disorders like Parkinson's disease (PD), Alzheimer's disease (AD), Huntington's disease (HD), multiple sclerosis (MS) and amyotrophic lateral sclerosis (ALS) alteration in blood-brain or blood-CSF transport is quite common (Bradbury, 1979; Hurtado-Alvarado, Domínguez-Salazar, Pavon, Velázquez-Moctezuma, & Gómez-González, 2016; Montagne et al., 2016; Montagne, Zhao, & Zlokovic, 2017; Ozkizilcik et al., 2018; Rapoport, 1976; Sharma, Castellani, Smith, & Sharma, 2012; Sharma et al., 2019; Sharma & Westman, 2004; Sweeney et al., 2019). This reflects an abnormal state of the BBB and BCSFB function (see Bradbury, 1979; Rapoport, 1976; Sharma & Westman, 2003, 2004). This dysfunction ultimately leads to pathologies abnormalities in the CNS in these diseases (Bosma et al., 2018; Erickson & Banks, 2018; Nation et al., 2019; Sharma et al., 2016; Sweeney, Zhao, et al., 2019).

The exact molecular of neurochemical basis of these neurological diseases, however, remains unknown. Available evidences indicate an involvement of the serotonergic system in several psychiatric diseases (Chakraborty et al., 2019; Conio et al., 2019; Daut & Fonken, 2019; Kraus, Castrén, Kasper, & Lanzenberger, 2017; Reiche et al., 2018; Shelton, 2019; Sinopoli, Burton, Kronenberg, & Arnold, 2017; Vadodaria, Stern, Marchetto, & Gage, 2018; Yohn, Gergues, & Samuels, 2017). In particular depression, emotional disturbances, schizophrenia and dementia altered levels of serotonin, its receptors, and metabolizing enzymes are well documented (see (Reiche et al., 2018; Shelton, 2019; Yohn et al., 2017). On the basis of either low or high serotonin syndromes in these psychiatric diseases drugs modifying serotonin metabolism are prescribed for clinical use in young or adult patients (Kraus et al., 2017; Shelton, 2019; Yohn et al., 2017).

In general, drugs that interfere with serotonin system or its metabolism in the CNS induce changes in psychological mood and behavior of the patients (Reiche et al., 2018; Shelton, 2019). In these clinical conditions of depression, anxiety or in mania cases, 5-HTP a precursor of serotonin is prescribed as a potent therapeutic agent (Birdsall, 1998; Fukuda, 2014, 2015; Jacobsen, Krystal, Krishnan, & Caron, 2016; Kaneko, Kumashiro, Takahashi, & Hoshino, 1979; Maes, Calabrese, Jayathilake, & Meltzer, 1997; Zmilacher, Battegay, & Gastpar, 1988). The drug is administered orally in doses varying from 50 to 150 mg/kg per day for extended periods that is continued the despite side effects involving disturbed mental function and abnormal behaviors (Jacobsen et al., 2016; Nakajima, Kudo, & Kaneko, 1978; Takahashi, Takahashi, Masumura, & Miike, 1976; van Vliet, Slaap, Westenberg, & Den Boer, 1996). Unfortunately, the mechanisms underlying 5-HTP induced abnormal behaviors and mental dysfunction are not well characterized (see Fahn, 1977; Martin, 1996; Sandyk, 2006).

Several lines of recent evidences indicate that serotonin is one of the powerful neurochemical agents that have the capacity to induce a reversible increase of the BBB permeability. This idea is further supported by the findings that application of serotonin on either the abluminal or luminal side of the cerebral microvessels lead to the opening of the BBB permeability. This effect of serotonin in the BBB permeability is mediated via specific 5-HT receptors (Sharma, 2000, 2004, 2007; Sharma & Dey, 1981; Sharma, Dey, & Olsson, 1989; Sharma, Olsson, & Dey, 1989, 1990a, 1995; Sharma, Westman, Cervós-Navarro, Dey, & Nyberg, 1995; Winkler, Sharma, Stålberg, Olsson, & Dey, 1995).

Furthermore, experimental conditions that invoke release of serotonin either following stress caused by immobilization, environmental heat stress or forced swimming, or that produced by mechanical trauma to the brain or spinal cord also elicit a breakdown of the BBB or blood-spinal cord barrier (BSCB) permeability (Dey & Sharma, 1983, 1984; Olsson, Sharma, Pettersson, & Cervos-Navarro, 1992; Sharma, 2004; Sharma & Cervós-Navarro, 1990; Sharma, Cervós-Navarro, & Dey, 1991a, 1991b; Sharma & Dey, 1981, 1984, 1986a, 1987, 1988; Sharma, Nyberg, Cervos-Navarro, & Dey, 1992; Sharma, Olsson, & Dey, 1990b; Sharma, Westman, Cervós-Navarro, et al., 1995; Sharma, Westman, Cervós-Navarro, Dey, & Nyberg, 1997; Sharma, Westman, Cervós-Navarro, & Nyberg, 1997; Sharma, Westman, & Nyberg, 1998; Sharma, Westman, Nyberg, Cervos-Navarro, & Dey, 1994). These observations suggest that elevation of endogenous serotonin levels influences the BBB Function (see Sharma, 2004; Sharma et al., 1990a).

Breakdown of the BBB to endogenous serum proteins into the brain fluid microenvironment lead of vasogenic edema formation (Sharma, 1999, 2004; Sharma et al., 1990b, 1992, 1994; Sharma, Dey, & Olsson, 1989). The magnitude and intensity of spread of edema fluid is one of the primary reasons for cell injury. Several lines of evidences suggest that exogenous elevation of serotonin in to the circulation causes disruption of the BBB, BCSFB and BSCB associated with brain edema formation.

5-HTP administrations induce significant increase in serotonin in the whole body, including plasma and the brain (Haberzettl, Bert, Fink, & Fox, 2013; Meltzer et al., 1982; Nakamura & Hasegawa, 2009; Udenfriend, Weissbach, & Bogdanski, 1957). Thus, it is quite likely that this treatment may also influence the BBB dysfunction, edema formation and cellular injuries (see Sharma, 2004). To our knowledge, there is no pervious information available on 5-HTP-induced BBB dysfunction. To expand our knowledge in this matter, the present investigation was undertaken to find out whether 5-HTP induced elevation of serotonin could induce a breakdown of the BBB permeability and induce edema formation associated with neuronal injures. Serotonin has powerful vasoconstrictor and/or vasodilation effects on the cerebral circulation (Bonvento, MacKenzie, Seylaz, & Lacombe, 1994; Edvinsson & MacKenzie, 1976; Kobayashi et al., 1985; Raskin, 1981). This effect of the amine is mediated through specific 5-HT receptors (Bonvento et al., 1994; Edvinsson & MacKenzie, 1976; Sharma, 2004; Sharma, Westman, Cervós-Navarro, et al., 1995). Thus, in this study the influence of 5-HTP on regional CBF was also investigated. Finally, to find out the receptor-mediated effect of serotonin on BBB function, brain edema and cellular injuries, several drugs modifying serotonin receptor or serotonin induced signal transduction mechanisms on the endothelial cell vesicular transports were also analyzed (Anon, 1989; Burnstock, 1985; MacKenzie & Scatton, 1987; Parsons, 1991; Saxena, Bom, & Verdouw, 1989; Wiernsperger, 1990). A brief description on our findings on 5-HT-induced neurotoxicity is discussed below.

2. Methodological considerations

2.1 Animals

Experiments were carried out on 164 inbred male Charles Foster Rats (110–150 g) housed at controlled room temperature (21 ± 1 °C) with 12 h light and 12 h dark schedule. Rat food pellets and tap water was provided ad libitum before the experiment.

2.2 Injection of 5-hydroxytryptophan

5-Hydroxytryptophan (5-hydroxy-L-tryptophan, Sigma Chemical Co., USA) was dissolved in sterile water and injected intraperitoneally (0.3 mL) in separate group of rats in a dose of 50 mg, 75 mg and 150 mg/kg. The animals were replaced in the cage after administration of the compound. The animals were kept alive 4 h after the injection.

2.2.1 Sham group

Saline treated animals (0.3 mL, i.p.) were used as sham group.

2.2.2 Control group

Normal intact rats served as controls.

2.3 Parameters measured

The following parameters were measured in control, saline treated and 5-HTP treated rats.

2.3.1 Stress and behavioral symptoms

Changes in rectal temperature were measured using a thermistor probe (Yellow Springfield, USA) connected to a 6-channel telethermometer (Aplab Electronics, USA). The other behavioral parameters such as salivation, apparent abnormal locomotion, jumping, circular motion and wet dog-shake activities were observed and scored (see Table 2). In addition, excretion of fecal pellets during the experiment and occurrence of gastric hemorrhages in the stomach wall was examined at autopsy (Sharma et al., 1991a; Sharma & Dey, 1986b, 1987).

2.3.2 Physiological variables

Changes in the mean arterial blood pressure (MABP), arterial pH, blood gases, heart rate and respiration were examined using standard procedures described earlier (Sharma, 2007; Sharma et al., 1990a; Sharma, Dey, & Olsson, 1989; Sharma, Olsson, & Dey, 1995; Sharma, Westman, Cervós-Navarro, et al., 1995; Winkler et al., 1995). The MABP was monitored from a polythene cannula (PE 25) inserted into the right common carotid artery and connected to a strain gauge pressure transducer (Statham P 23, USA). The out-put from the transducer was connected to a chart recorder (Electromed, UK). Specific leads placed on the abdominal skin over the heart and a respiratory belt tied over the belly were used to record heart rate and respiration, respectively (Harvard Apparatus, USA). The outputs were also recorded in the chart recorder (Electromed, UK) (Sharma, Ali, Hussain, Schlager, & Sharma, 2009; Sharma et al., 2009).

At the time of connecting the arterial cannula to the pressure transducer, a sample of arterial blood was withdrawn for the measurement of the arterial pH, PaO_2 and $PaCO_2$ using a Radiometer apparatus (Copenhagen) (Sharma, 1987).

2.4 Blood-brain barrier permeability

Four hours after saline or 5-HTP administration, animals were anesthetized with urethane (1 g/kg, i.p.). Evans blue (2% of a 0.5 mL solution) containing 10 μCi $^{[131]}$I-sodium (about 1 million counts per minute) were administered into the right femoral vein through a needle puncture (Sharma, 2007; Sharma, Dey, & Olsson, 1989; Sharma et al., 1990a; Sharma, Westman, Cervós-Navarro, et al., 1995; Winkler et al., 1995). The tracer was allowed to circulate for 5 min. At the end of the experiment, the intravascular tracer was washed-out with a brief saline rinse via heart (45 s). Immediately before perfusion 1 mL arterial blood sample was withdrawn through a needle puncture from the left ventricle to determine whole blood radioactivity at the time of killing (Sharma, 2007; Sharma & Dey, 1986a, 1986b, 1987, 1988).

After perfusion, the brain was removed and bisected in the mid line. In one half of the brain extravasation of Evans blue dye was examined using a magnifying lens. The dye entered into the brain was measured calorimetrically as described earlier (Sharma, 1987; Sharma, Ali, Hussain, et al., 2009; Sharma, Ali, Tian, et al., 2009). The other half brain was dissected into the 14 brain regions that were weighed immediately and counted for radioactivity in a three-in well type gamma counter (Beckman, Germany). The extravasation of radioiodine was expressed as percentage over the blood radioactivity (see Sharma, 1987, 2007; Sharma & Dey, 1981; Sharma, Dey, & Olsson, 1989; Sharma, Olsson, & Dey, 1989, 1995; Sharma, Westman, Cervós-Navarro, et al., 1995; Winkler et al., 1995).

2.5 Cerebral blood flow

The CBF was measured once in different brain regions using carbonized tracer microspheres (OD 15 μm ± 0.6 μm) labeled to $^{[125]}$Iodine as described earlier (Sharma, 1987; Sharma et al., 1990b, 1992; Sharma & Dey, 1986b, 1987; Sharma, Dey, & Olsson, 1989; Sharma, Olsson, & Dey, 1989; Sharma, Westman, Cervós-Navarro, Dey, & Nyberg, 1997). In brief, about 10^6 microspheres were injected near the left atrium through a cannula (PE 25) implanted into the left common carotid artery. The tip of the cannula was advanced retrogradely toward the heart (see Sharma, 1987).

Timed arterial blood samples at every 30 s from the periphery were collected from the right femoral artery at the rate of 0.8 mL/min starting from 30 s before infusion and continued until 90 s after the infusion (Sharma, 1987; Sharma et al., 1990b; Sharma & Dey, 1986b, 1987) (Figs. 1 and 2).

After 90 s of bolus injection, the animals were decapitated and the brain and spinal cord was removed immediately and placed on filter paper wetted with ice-cold saline. The large superficial vessels were removed and discarded. The brain and spinal cord was then divided into 14 anatomical regions (Fig. 3), weighed immediately and counted in a three-in well type Beckman Gamma Counter (USA) at 50–80 keV energy window (Sharma, 1987; Sharma et al., 1990b; Sharma & Dey, 1986b, 1987).

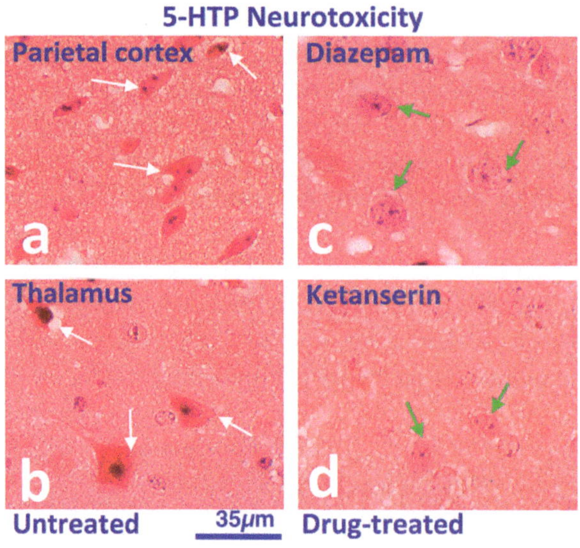

Fig. 1 High power light micrograph of Hematoxylin and Eosin (H&E) stained paraffin sections (3-μm thick) showing neuronal damages in 5-HTP (150 mg/kg, i.p.) treated (a and b) rat and neuroprotection induced by diazepam (c) and ketanserin (d). Several neurons (white arrows) showed distortion, perineuronal edema and sponginess in the neuropil of parietal cerebral cortex (a) and thalamus (b). Dark neuronal cell cytoplasm and karyoplasm in some neurons are clearly visible (a and b). Pretreatment with diazepam (c) or ketanserin (d) markedly attenuated neuronal damages in 5-HTP treated groups. Several neurons (green arrows) show clear neuronal cytoplasm and karyoplasm with compact neuropil. Perineuronal edema is much less distinct in drug-treated group after 5-HTP-administration. For details see text. Bar = 35 μm.

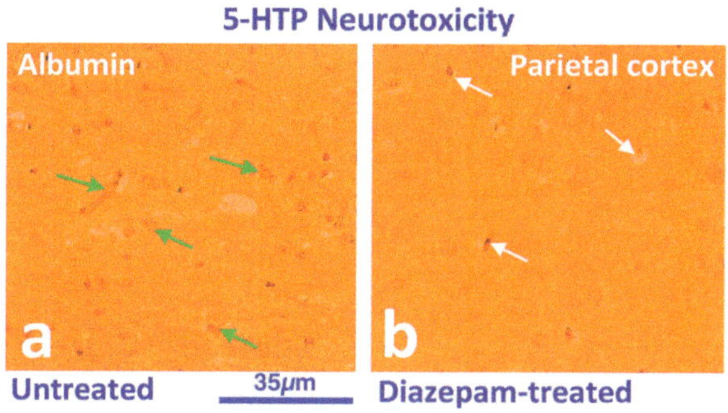

Fig. 2 Low power light micrograph of albumin immunostaining on 3-μm thick paraffin sections of rat brain parietal cerebral cortex after 4h of 5-HTP (150mg/kg, i.p.) treated (a) its modification with diazepam pretreatment (b). Albumin extravasation (green arrows) is clearly evident in the neuropil around microvessel and neurons. Perineuronal vacuolation showing albumin deposit is clearly visible (a). Pretreatment with diazepam (b) markedly reduced albumin leakage in 5-HTP administered rat. Few albumin positive cells (white arrows) are only visible in 5-HTP intoxicated diazepam treated group (b). Bar=35μm.

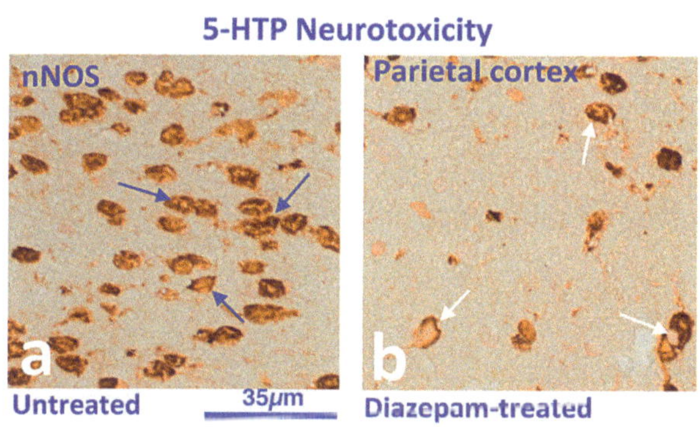

Fig. 3 Low power light micrograph of neuronal nitric oxide synthase (nNOS) immunoreactivity in the parietal cerebral cortex in the rat following 4h after 5-HTP (150mg/kg, i.p.) administration (a) and its reduction in diazepam pretreated (b) group. Several nNOS positive neurons could be seen (blue arrows) in the 5-HTP treated groups in the neuropil. Diazepam pretreatment in 5-TP intoxicated animal markedly reduced the nNOS upregulation (b). Only a few nNOS positive cells could be seen in diazepam pretreated 5-HTP group (white arrows). Paraffin sections 3-μm thick, Bar=35μm. For details see text.

The CBF (mL/g/min) was calculated according to the formula: CPM Brain Tissue/Total counts in reference blood flow × 0.8 (reference blood flow) (Heistad, Marcus, & Mueller, 1977; Marcus, Busija, Bischof, & Heistad, 1981; Marcus, Heistad, Ehrhardt, & Abboud, 1977; Sharma, 1987).

2.6 Brain edema formation and volume swelling

After counting the radioactivity for measuring regional CBF, the tissue samples that are already weighed (wet weight) on a preweighed filter paper were placed in an oven maintained at 90 °C for 72 h to evaporate the water content of the tissue (Kiyatkin, Brown, & Sharma, 2007; Kiyatkin & Sharma, 2009; Sharma, 2007; Sharma et al., 1992, 1994, 2015; Sharma & Kiyatkin, 2009; Sharma, Westman, Cervós-Navarro, et al., 1995). After that, the dry weight of the samples was recorded on the same microbalance at three different occasions until the dry weight become constant. The brain water content representing edema formation was calculated in control, experimental and drug treated animals using the formula: Wet weight—dry weight/wet weight × 100 (Olsson et al., 1992; Sharma et al., 1991b; Sharma & Cervós-Navarro, 1990; Sharma & Olsson, 1990; Sharma, Olsson, & Cervós-Navarro, 1993).

Volume swelling from the changes in brain water between control and experimental group were calculated suing formula of Elliott and Jasper (1949). Roughly, 1% increase in brain water is equal to 4% increase in volume swelling (Dey & Sharma, 1983, 1984; Kiyatkin et al., 2007; Olsson et al., 1992; Sharma et al., 1991b, 1993, 1994; Sharma, Ali, Hussain, et al., 2009; Sharma & Olsson, 1990; Sharma, Westman, & Nyberg, 1998).

2.7 Measurement of serotonin

Separate group of rats were anesthetized with urethane 4 h after saline or 5-HTP treatment. Simultaneous measurement of 5-HT and the BBB permeability was not carried out in these animals due to obvious reasons that injection of Evans blue and brain perfusion may have interfered with the biochemical determination (Sharma et al., 1990a; Sharma, Westman, Cervós-Navarro, et al., 1995).

We measured serotonin in plasma and in brain using a sensitive and specific spectrophotofluorometric assay (Sharma, 1982; Sharma et al., 1991a; Sharma & Dey, 1981, 1986a, 1987, 1988; Sharma, Westman, Cervós-Navarro, et al., 1995; Snyder, Axelrod, & Zweig, 1965). In rats after experiments, about 1 mL of whole blood was withdrawn from the left ventricle and centrifuged immediately to obtain plasma. Plasma (0.5 mL) was diluted with perchloric acid (PCA) and protein was precipitated out, centrifuged, and the supernatant was collected (Sharma & Dey, 1986a, 1987, 1988). The brain was removed after decapitation and placed in ice cooled 0.4 N perchloric acid and stored at 4 °C overnight (Sharma, 1982; Sharma & Dey, 1981). On the second day, the brain was dissected into eight regions (see Table 2). Each region was homogenized in PCA and centrifuged at room temperature (×900 g) and the supernatant was obtained (Sharma & Dey, 1981). 1 mL of aliquots from plasma and brain homogenates was used for serotonin measurement. The extraction of serotonin from these plasma and brain samples was carried out in butanol (extra pure grade, BDH, London, UK) in a salt saturated alkaline medium (pH 10). The purification of serotonin extract was done using *n*-heptane (Extra pure, Merck, Germany) (Snyder et al., 1965).

The fluorspars were develop by incubating the samples with ninhydrin (Sigma Chemical Co., St. Louis, USA) at 75 °C for 30 min (Sharma & Dey, 1986a, 1987, 1988; Snyder et al., 1965). The florescence was measured in triplicate in cooled samples at room temperature (21 ± 1 °C) in an AMINCO-Bowman Spectrophotofluorometer (Waltham, MA, USA) at excitation 385 nm and emission 490 nm wavelengths as described earlier (Sharma et al., 1991a; Sharma & Dey, 1986a, 1987, 1988; Sharma, Westman, Cervós-Navarro, et al., 1995; Snyder et al., 1965).

2.8 Brain pathology

In separate group of control, 5-HTP intoxicated and drug-treated groups after experiments were deeply anesthetized with urethane (1 mg/kg, i.p.) and perfused with 4% buffered paraformaldehyde after brief saline rinse (Sharma, 2007; Sharma et al., 1991b, 1993; Sharma & Cervós-Navarro, 1990; Sharma & Olsson, 1990). For this purpose, after anesthesia, the chest was rapidly opened and heart was exposed. The right auricle was excised and a perfusion butterfly cannula was inserted into the left ventricle connected with bottle filled with 0.9% sterile cold saline (4 °C) (see Sharma, 2007). About 50 mL of saline was perfused through heart to clear remaining blood from the circulation system. After the perfusion system was switched to bottle filled with cold paraformaldehyde in 0.1 M phosphate buffered saline (pH7.0). About 250 mL fixative was perfused through heart at 90 mm Torr perfusion pressure (Sharma et al., 1993; Sharma & Olsson, 1990).

After completion of the perfusion, the animals were wrapped in aluminum foil and kept in a refrigerator at 4 °C overnight (Sharma et al., 1993; Sharma & Olsson, 1990). On the next day, the brain and spinal cord was dissected out and desired pieces from different areas were cut and placed in the same fixative at 4 °C for 1 week. After that coronal sections of the brain and spinal cord were embedded in paraffin using standard protocol and 3-μm thick sections were cut and processed for histological or immunohistochemical staining (see below) (Sharma et al., 1991b; Sharma & Cervós-Navarro, 1990; Sharma & Olsson, 1990).

2.8.1 Light microscopy

Neuronal injury and neuropil were examined using Nissl or Hematoxylin and Eosin (H&E) staining on 3-μm thick paraffin section using standard protocol (Sharma et al., 1993; Sharma & Olsson, 1990). The stained sections were examined under a Zeiss Bright field upright Microscope (Axio Imager. Z2m, Germany). The images were captured on a digital camera attached to it and processed in a Apple Macintosh Computer Mac OX X using software Adobe Photoshop 10.2 using identical color filters from various groups for comparison (Pettersson, Sharma, & Olsson, 1990; Sharma, Gordh, Wiklund, Mohanty, & Sjöquist, 2006; Sharma & Sjöquist, 2002).

2.8.2 Transmission electron microscopy

For ultrastructural analysis of vascular permeability using lanthanum-an electron dense tracer (molecular diameter 12Å) was examined using standard protocol (see Olsson et al., 1992; Pettersson et al., 1990; Sharma et al., 2006; Sharma, Olsson, Persson, & Nyberg, 1995). For this purpose, in separate group of rats after experiments, were perfused with fixative containing lanthanum as described earlier. In brief, the lanthanum chloride (2.5%) was dissolved in 0.4% paraformaldehyde in 0.1M phosphate buffer saline (pH 7.0) containing 0.2% picric acid and about 200mL of this fixative was perfused in situ via cardiac puncture after rinsing the microvessels with 0.9% buffered saline (4°C, 50mL) at 90Torr perfusion pressure (Olsson et al., 1992; Pettersson et al., 1990; Sharma et al., 2006).

Perfusion with lanthanum has advantages to examine the passage of the tracer across the endothelial cells at ultrastructural level using transmission electron microscopy (TEM) as black particles with without any specific processing (Pettersson et al., 1990; Sharma et al., 2006; Sharma, Olsson, Persson, & Nyberg, 1995). Thus, after perfusion, small tissue pieces were postfixed in OSO_4 and embedded in Epon 812. Semithin sections (about 1 μm) were cut and stained with toluidine blue staining to identify the area for ultrathin sections. Ultrathin sections were cut using diamond knife at LKB Ultramicrotome (Stockholm, Sweden) and serial sections were collected on a one hole copper grid (Sharma, 2007; Sharma, Westman, & Nyberg, 1998). Some sections were examined unstained whereas; some sections were stained for contrast with lead citrate and uranyl acetate and examined under a Phillips 400 TEM. The images were captured on a digital camera and processed using identical filters in a Macintosh computer using Adobe Photoshop software 10.2 (Olsson et al., 1992; Pettersson et al., 1990; Sharma et al., 2006; Sharma, Olsson, Persson, & Nyberg, 1995).

2.8.3 Immunohistochemistry for albumin and neuronal nitric oxide synthase

Immunohistochemistry (IHC) was performed on 3-μm thick paraffin sections for endogenous albumin leakage and neuronal nitric oxide synthase (nNOS) according to standard protocol (Gordh, Chu, & Sharma, 2006; Sharma et al., 1998, 2015; Sharma, Alm, & Westman, 1998; Sharma, Badgaiyan, Alm, Mohanty, & Wiklund, 2005; Sharma, Miclescu, & Wiklund, 2011; Sharma, Westman, Olsson, & Alm, 1996).

(a) *Albumin immunohistochemistry*

IHC for albumin was done on paraffin sections using a sheep polyclonal ant-rat albumin antibody (Sigma, USA) and the streptavidin-horse radish peroxidase (HRP)-biotin techniques (Gordh et al., 2006; Sharma et al., 2011; Sharma, Patnaik, et al., 2015). In brief, the endogenous peroxidase activity was blocked with 3% H_2O_2 and 5% normal goat serum. After that, the sections were incubated with the primary antibodies (1:500) for overnight with shaking at 4 °C. This was followed by incubation with biotinylated linking antibody and HRP (DAKO, Germany) with brief rinses in phosphate buffer saline (PNS) between incubations. The reaction was visualized using 3-amino-9-ethylcarbazole (Vector Lab, Burlingame, USA) and some of the sections were counterstained with Hematoxylin and Eosin (HE). For technical control sections were omitted with primary antibody incubation or substituted with nonimmune serum for the primary antibody in the staining protocol simultaneously to confirm the specificity of the primary antibody used (Gordh et al., 2006; Sharma et al., 2011).

The number of albumin positive cells was counted in blinded fashion by three observers and median values are used for calculations in various groups.

(b) *Neuronal nitric oxide synthase (nNOS) immunohistochemistry*

The nNOS IHC was dome on 3-μm thick paraffin sections using commercial nNOS antiserum (1:5000 in PBS) using DAB technique (Sharma et al., 1996, 2005; Sharma, Alm, & Westman, 1998). In brief, the sections were incubated with primary nNOS antibodies under agitation at room temperature for 36 h together with normal swine serum (1:30 in PBS). After that the sections were rinsed 10 min six times in PBS and incubated with swine antirabbit immunoglobulin (1:30 in PBS) for 60 min at room temperature with several rinses in PBS. Thereafter, the sections were incubated with rabbit peroxidase-antiperoxidase complex (PAP) (1:20 in PBS) and rinsed several times with PBS (see Sharma et al., 1996, 2005; Sharma, Alm, & Westman, 1998).

The immunoreactivity was developed on sections by incubating them for 6–7 min in a solution containing 75 mg of DAB and 30 μL of 30% H_2O_2/100 mL of Tris–HCl buffer. After that the sections were washed with sodium cacodylate buffer and processed for light microscopy using commercial protocol (Sharma et al., 1996). The number of nNOS positive cells were counted and identified for comparison as mentioned above.

3. Pharmacological treatments

In separate groups of rats, the following drugs were given as pre-treatment (Table 2). In these drug treated rats, BBB permeability, CBF, 5-HT, brain edema and cell injury were examined according to the standard protocol.

(a) *p-Chlorophenylalanine (a 5-HT synthesis inhibitor)*: *p*-CPA was dissolved in sterile saline and administered intraperitoneally in animals (100 mg/kg) daily for 3 days. On the 4th day, animals received 5-HTP injections. This dose and time schedule depletes serotonin content of blood as well and in the CNS for long-time (Andén & Modigh, 1972; Bhagavan & Coursin, 1973; Sicuteri, Anselmi, & Fanciullacci, 1970).

(b) *Indomethacin (a prostaglandin synthetase inhibitor)*: The drug was dissolved in saline with addition of a pinch of sodium bicarbonate at 60 °C. The pH was adjusted to 7.4 before administration of indomethacin (10 mg/kg) in animals through intraperitoneal route. These animals received 5-HTP injection 30 min after indomethacin treatment. This dose of the drug induces inhibition of prostaglandins in the CNS for several hours (Crook & Collins, 1977; Gafni, Schwartzman, & Raz, 1978; Sharpe, Thalme, & Larsson, 1974).

(c) *Diazepam (an anxiolytic drug)*: Diazepam ampoule was diluted in saline and injected subcutaneously (4 mg/kg) 30 min before 5-HTP administrations. This dose of the compound is known induce anxiolytic effects in rats (Dailly, Hascoët, Colombel, Jolliet, & Bourin, 2002; Kerry & Jenner, 1962; Yeung, Treit, & Dickson, 2012).

(d) *Ketanserin (a 5-HT2 receptor antagonist)*: The drug was dissolved in 30% ethanol and administered 30 min (1 mg/kg, i.p.) before 5-HTP administration. This dose of the drug effectively blocks 5-HT2 receptors in the CNS (de Clerck, David, & Janssen, 1982; Leysen et al., 1981; Van Nueten, Xhonneux, Vanhoutte, & Janssen, 1981).

(e) *Cyproheptadine (a 5-HT2-receptor antagonist with weak effects on Histamine H1 receptors)*: The drug was dissolved in saline and injected intraperitoneally (5 mg/kg) 30 min before 5-HTP administrations. This dose of compound mainly inhibits 5-HT2 receptors. The pre synaptic 5-HT receptors or auto receptors are not blocked with this dosage used (Doggrell, 1992; Krstić & Katusić, 1982; Reiche & Frey, 1983).

(f) *Vinblastine (an inhibitor of vesicular transport)*: Vinblastine (0.8 mg/kg) was administered intravenously 30 h before 5-HTP challenge. The drug in this dose is known to inhibit mitosis in microtubules thereby effectively able to block formation of channels or vesicular transport in endothelial cells (Dustin, 1963; Florian & Mitchison, 2016; Johnson, Armstrong, Gorman, & Burnett, 1963).

3.1 Statistical analysis

The unpaired student's t-test was applied to evaluate statistical significance between control and experimental groups. ANOVA followed by Dunnett's test was applied for multiple comparison involving drug-treatments. A *P* value <0.05 was considered to be significant.

4. Our observations on 5-HTP neurotoxicity

4.1 Effect of 5-HTP on stress and behavioral symptoms

Administration of 5-HTP in conscious rats resulted in development of stress symptoms, which is dose dependent (Table 1). At the end of 4 h after 5-HTP injections, animals exhibited a dose dependent hypothermia with 50 mg (-0.54 ± 0.04 °C) and 75 mg (-1.18 ± 0.02 °C) respectively. However, with further increase in doe to 150 mg, significant hyperthermia ($+3.46 \pm 0.23$ °C, $P < 0.001$, compared to control) was noted. The hyperthermic animals exhibited signs of salivation, a feature not observed in animals, which received lower dosage of the compound. Animals with 5-HTP 150 mg dose showed hyperlocomotor activity compared to untreated or saline treated rats. This effect was dose dependent. The other behavioral symptoms such as jumping, wet dog shake and circular motion exhibited a dose dependent response compared to normal animals (Table 1). Animals received highest dose of 5-HTP showed diarrheic stool. Postmortem examination revealed microhemorrhages in the stomach wall in animals with 150 mg dose of 5-HTP.

Table 1 Effect of 5-hydroxytryptophan on stress symptoms and physiological variables.

Parameters measured	Control	Saline	5-HTP treatment		
			50 mg/kg	75 mg/kg	150 mg/kg
Stress symptoms	$n=6$	$n=5$	$n=5$	$n=6$	$n=8$
Δ°C Rectal T	+0.40 ± 0.30	+0.44 ± 0.21	−0.54 ± 0.04[b]	−1.18 ± 0.02[a]	+3.46 ± 0.23[a]
Salivation	−	−	−	−	+++
Locomotion	+	+	++	++	++++
Defecation pellets (total)	5 ± 2	4 ± 1	8 ± 2	14 ± 4[a]	20 ± 6#[a]
Wet dog shake	−	−	−	4 ± 2	8 ± 3
Jumping	−	−	+/−	++	+++
Circular motion	−	−	−	+/−	++
Grooming	−	−	+	+	++
Gastric hemorrhages	−	−	−	4 ± 2	20 ± 4[c]
Physiological variables	$n=6$	$n=5$	$n=5$	$n=5$	$n=5$
MABP torr	104 ± 6	108 ± 4	120 ± 8[b]	128 ± 6[b]	82 ± 8[a]
Arterial pH	7.38 ± 0.04	7.38 ± 0.06	7.37 ± 0.04	7.36 ± 0.08	7.36 ± 0.05
Arterial PaO_2 torr	80.23 ± 0.21	80.24 ± 0.18	79.25 ± 0.34	78.34 ± 0.56	81.23 ± 01.6
Arterial $paCO_2$ torr	34.36 ± 0.23	34.56 ± 0.43	33.28 ± 0.18	33.23 ± 0.23	36.13 ± 1.15
Heart rate beats/min	64 ± 4	66 ± 3	68 ± 5	70 ± 6	74 ± 4
Respiration cycle/min	35 ± 5	33 ± 4	38 ± 4	40 ± 6	44 ± 6

a = significantly different from control; b = significantly different from 5-HTP 150 mg dose; c = significantly different from 5-HTP 75 mg dose. # In few cases ($n=3$) partial diarrheic stool; − = absent, + = very mild, ++ = mild, +++ = moderate, ++++ = extensive; * = $P<0.05$, *** = $P<0.001$, Student's unpaired t-test.

4.2 Effect of 5-HTP on physiological variables

Administration of 5-HTP induced widespread changes in physiological variables (Table 1). Lower dosage of 5-HTP (50 and 75 mg) induced a mild hypertension while 150 mg dose of the compound resulted in hypotension (20 Torr). The arterial pH did not show significant difference from the control group irrespective of the dose used. There was a trend of increase PaO_2 as well $PaCO_2$ at the end of 4 h after 5-HTP 150 mg administration. However, this difference was not statistically significant. With low dose of the compound, a slight insignificant decrease in PaO_2 and $PaCO_2$ were noted. 5-HTP treatments on the other hand resulted in dose dependent changes in heart rate. Low dose of the compound induced bradycardia whereas, 75 and 150 mg dose induced tachycardia compared to saline treated animals. Low dose of 5-HTP resulted in slight hypopnea, 75 and 150 mg dose resulted in hyperpnoea (Table 1).

4.3 Effect of 5-HTP on blood-brain barrier permeability

Administration of 150 mg dose of 5-HTP significantly increased the permeability of the BBB to Evans blue and $^{[131]}$Iodine compared to normal control and sham treated group (Table 1). Evans blue dye was seen in the cerebral cortex superficial layers as well as on the dorsal surface of the hippocampus and caudate nucleus. The cerebellar vermis and occasionally lateral cerebellar cortex were also stained. Extravasation of radioiodine was further extended to many other regions (Table 3). Lower doses of the compound did not induce extravasation of either tracer.

In general, the magnitude and severity of radiotracer extravasation was much higher than the Evans blue dye permeability.

This increase in the permeability of the BBB following 5-HTP (150 mg) dose was reversible in nature. Thus, in separate group of rats, when the tracer substances were administered 10 h after the 5-HTP administration, no significant increase in the permeability to either Evans blue or radio-iodine was noted (Table 2).

Regional BBB permeability showed wide range of variation in the magnitude of radiotracer extravasation. Thus cortical regions exhibited a minimum of 175% increase (frontal cortex) to a maximum of 722% higher (cingulate cortex) permeation of radioiodine compared from controls. The cerebellar cortices showed even further high extravasation (777–925% increase from the control). Hippocampus, caudate nucleus and brain stem showed 485–731% increase in tracer extravasation. On the other hand,

Table 2 Effect of 5-hydroxytryptophan on the blood-brain barrier permeability and cerebral blood flow.

			BBB permeability		Cerebral blood flow
Type of experiment	***n***	**Rectal T (°C)**	**EBA mg%**	**$^{[131]}$Iodine %**	**Whole brain mL/g/min**
Control group	6	36.83±0.23	0.24±0.04	0.34±0.03	1.12±0.04
Saline	5	37.04±0.18	0.22±0.06	0.38±0.06	1.03±0.05
		+0.21	−8.3	+10.5	−8.03
5-HTP 50 mg/kg	5	36.01±0.04*	0.28±0.10	0.42±0.10	1.34±0.08*
		−0.82	+16.67	+23.52	+19.64
5-HTP 75 mg/kg	6	35.67±0.23*	0.26±0.08	0.38±0.06	0.94±0.06
		−1.16	+8.33	+11.76	−16.07
5-HTP 150 mg/kg	8	40.48±0.24***	1.34±0.65***	2.34±0.65***	0.76±0.10***
		+3.65	+458.33	+588	−32.14
5-HTP 150 mg/kg[a]	6	37.64±0.34	0.54±0.23	0.68±0.28	0.96±0.08
		+0.81	+58.82	+100	−14.28

[a]Values measured 10 h after 5-HTP injection.

Animals received intraperitoneal injection of saline or 5-HTP in sterile water (0.3 mL). Four hours after administration of drug or saline body temperature, BBB permeability and CBF were measured.

Values are mean ± SD; Figures below the mean values indicate % change from control. − = below, + = above the control values. BBB = blood-brain barrier; CBF = cerebral blood flow.

* = $P < 0.05$, *** = $P < 0.001$, Student's unpaired t-test.

Table 3 Effect of 5-HTP on plasma and regional brain serotonin levels.

Serotonin content	Control	Saline	5-HTP treatment		
			50 mg/kg	75 mg/kg	150 mg/kg
Brain serotonin (μg/g)	$n=6$	$n=5$	$n=5$	$n=6$	$n=8$
Cerebral cortex	0.67 ± 0.21	0.72 ± 0.43	0.56 ± 0.23	0.89 ± 0.32[a]	2.36 ± 1.23[b]
		−7.46	−16.41	+32.83	+525.72
Hippocampus	0.34 ± 0.15	0.38 ± 0.08	0.89 ± 0.21[a]	1.15 ± 0.34[a]	1.33 ± 0.34[a]
		+11.76	+161.76	+238.23	+291.17
Cerebellum	0.28 ± 0.07	0.33 ± 0.06	0.56 ± 0.08[a]	0.89 ± 0.08[a]	0.95 ± 0.06[a]
		+17.85	+100	+217.85	+239.28
Thalamus	0.54 ± 0.12	0.48 ± 0.18	0.38 ± 0.17[b]	0.86 ± 0.17	1.67 ± 0.23[a]
		−11.11	−29.62	+59.25	+209.25
Hypothalamus	0.43 ± 0.12	0.46 ± 0.16	0.67 ± 0.14[b]	0.76 ± 0.18[b]	1.87 ± 0.23[a]
		+6.97	+55.81	+76.74	+334.88
Medulla	0.35 ± 0.11	0.36 ± 0.08	0.66 ± 0.12[b]	0.68 ± 0.14[b]	1.24 ± 0.18[a]
		+2.85	+88.57	+94.28	+254.28
Brain stem	0.48 ± 0.08	0.54 ± 0.10	0.68 ± 0.12[b]	0.76 ± 0.08[a]	0.89 ± 0.14[a]
		+12.5	+41.66	+58.33	+85.41
Plasma μg/mL	0.24 ± 0.04	0.26 ± 0.03	0.34 ± 0.06[a]	0.58 ± 0.08[a]	0.87 ± 0.10[a]
		+8.33	+41.66	+141.66	+262.5

[a] $P < 0.001$.
[b] $P < 0.05$.
Values are mean ± SD; Figures below mean values indicate % change from control group. − = below, + = above the control value.
Student's unpaired t-test compared from control group.

thalamus, hypothalamus, pons, medulla and frontal cortex exhibited only 175–293% increase. The inferior colliculus showed an insignificant increase of 7% only (Table 3).

4.4 Effect of 5-HTP on cerebral blood flow

Subjection of animals to 4 h after 5-HTP injection (150 mg) resulted in a marked 32% decrease in the cerebral blood flow. This decrease in CBF was not seen after 10h 5-HTP administration. Thus, at this time period, the CBF showed a nonsignificant decline by 16% only (Table 2). Administration of 50mg dose in fact increased the CBF by 20% after 4h. The 75mg dose did not significantly influence the CBF changes.

Analysis of regional CBF showed widespread reduction in flow in 150mg treated rats at 4h. The cerebral and cerebellar cortices, hippocampus, caudate nucleus showed 21–24% decline in the CBF. Thalamus, hypothalamus, pons, medulla and brain stem showed a 2–17.5% decline in the regional CBF. Whereas, colliculi exhibited 8–13% decrease in the CBF compared to control group.

The magnitude and severity of CBF reduction were in no way comparable to the increase in the regional BBB permeability.

4.5 Effects of 5-HTP on brain edema formation and volume swelling

Regional brain edema was measured in eight brain regions (Table 4) after saline or 5-HTP administration in selected doses. Our observations showed a dose related increase in the brain water content and volume swelling in 5-HTP treated rats. The cerebral cortex showed 3%, 5% and 11% increase in volume swelling after 5-HTP 50, 75 and 150mg doses, respectively. Interesting cerebellum was the top region that showed highest volume swelling that was higher than the cerebral cortex in 5-HTP treated rats. Thus, 4%, 9% and 11% volume swelling was seen in the cerebellum after 5-HTP 50, 75 and 150mg doses, respectively. The other brain regions namely hippocampus, thalamus, hypothalamus, medulla and brain stem showed volume swelling of 4%, 6% and 7–8% after 5-HTP administrations in doses 50, 75 and 150mg (Table 4).

4.6 Effect of 5-HTP on plasma and brain 5-HT concentration

We measured 5-HT concentration in the plasma and brain 4h after 5-HTP administrations. The plasma 5-HT showed a dose dependent increase from 41% to 262.5% following 50 to 150mg doses of the 5-HT precursor. In brain there is a regional variation in serotonin concentration after 5-HTP administrations. In 50mg group, hippocampus and cerebellum showed 161% and 100% increase in 5-HT concentration, respectively. Hypothalamus, medulla and brain stem exhibited about 41% to 88% increase in the amine level. The cerebral cortex and thalamus in fact showed a mild 16–22% decreases in the 5-HT level.

Table 4 Effect of 5-HTP on brain water content and volume swelling.

Brain edema	Control	Saline	5-HTP treatment 50 mg/kg	75 mg/kg	150 mg/kg
Brain water %	$n=6$	$n=5$	$n=5$	$n=6$	$n=8$
Cerebral cortex	74.34 ± 0.21	74.56 ± 0.32	75.12 ± 0.13[a]	75.67 ± 0.21[a]	77.12 ± 0.35[b]
			+3.12	+5.32	+11.12
Hippocampus	68.34 ± 0.12	68.53 ± 0.09	69.34 ± 0.21[a]	69.98 ± 0.11[a]	70.56 ± 0.21[b]
			+4.00	+6.56	+8.96
Cerebellum	71.23 ± 0.06	71.32 ± 0.10	72.34 ± 0.22[b]	73.41 ± 0.22[b]	74.10 ± 0.21[b]
			+4.44	+8.72	+11.48
Thalamus	67.34 ± 0.32	67.28 ± 0.18	68.23 ± 0.16[a]	68.97 ± 0.23[b]	69.04 ± 0.14[b]
			+3.56	+6.52	+6.80
Hypothalamus	65.23 ± 0.08	65.67 ± 0.21	66.23 ± 0.16[b]	66.74 ± 0.21[b]	67.34 ± 0.18[b]
			+4.00	+6.04	+8.44
Medulla	65.34 ± 0.21	65.67 ± 0.31	66.28 ± 0.25[b]	66.87 ± 0.29[b]	67.23 ± 0.30[b]
			+3.76	+6.12	+7.56
Brain stem	64.39 ± 0.25	64.45 ± 0.52	65.32 ± 0.25[b]	65.87 ± 0.12[b]	66.46 ± 0.15[b]
			+3.73	+5.92	+8.28

[a] $P < 0.05$.
[b] $P < 0.01$.
Values are Mean ± SD; Figures below mean values indicate % *f* (volume swelling) from control group. − = below, + = above the control value. Student's unpaired t-test compared from control group.

With 75 mg dose, all brain regions showed a significant increase. The most pronounced increase (two- to threefold) was seen in the hippocampus and cerebellum. The other regions exhibited 30–100% increase. With further increase in dose to 150 mg, the cerebral cortex showed fivefold increase in 5-HT level. Thalamus, hypothalamus and medulla also exhibited two- to threefold increase of the amine. Hippocampus and cerebellum was still two- to threefold higher in their 5-HT content. Brain stem showed only 85% increase in the amine level.

4.7 Effects of 5-HTP on brain pathology

Neuronal injuries were examined in 5-HTP treated rats using Nissl or H&E staining on paraffin sections (Table 5). In general, neuronal injuries were seen in almost all eight-brain regions examined in a dose dependent manner. Interestingly the number of damaged neurons was highest in the cerebellum followed by hippocampus and cerebral cortex (Table 5). Cerebellum showed mean 34, 44 and 63 damaged neurons in the cerebellar cortical areas following 5-HTP administrations of 50, 75 and 150 mg doses. On the other hand hippocampus exhibited 54, 78 and 89 damaged neurons in the area

Table 5 Effect of 5-HTP on neuronal injury using Nissl or Hematoxylin and Eosin (H&E) staining on 3-μm thick paraffin sections.

			5-HTP treatment		
Neuronal Injury Nrs.	**Control**	**Saline**	**50 mg/kg**	**75 mg/kg**	**150 mg/kg**
Neuronal injury	$n=6$	$n=5$	$n=5$	$n=6$	$n=8$
Cerebral cortex	3 ± 4	2 ± 3	24 ± 6^a	46 ± 8^a	78 ± 8^a
Hippocampus	2 ± 1	2 ± 3	34 ± 8^a	44 ± 10^a	63 ± 12^a
Cerebellum	3 ± 2	3 ± 4	54 ± 12^a	78 ± 14^a	89 ± 18^a
Thalamus	4 ± 2	4 ± 5	18 ± 5	26 ± 8	43 ± 12^a
Hypothalamus	2 ± 1	2 ± 2	12 ± 6^a	18 ± 8^a	34 ± 7^a
Medulla	3 ± 1	3 ± 2	8 ± 2^b	12 ± 6^b	20 ± 4^a
Brain stem	2 ± 2	2 ± 3	6 ± 2^b	8 ± 3^b	13 ± 4^b

$^aP<0.01$.
$^bP<0.05$.
Neuronal injury is characterized by deform, swollen or shrunken neurons associated with perineuronal edema, loss of nucleolus or eccentric nucleolus within the nucleus of nerve cells in anatomical defined areas. The number of such neurons is examined in a blinded fashion by three observers and median values are used for statistical analysis.
Values are Mean ± SD; Student's unpaired t-test compared from control group.

following 50, 75 and 150 mf doses of 5-HTP within 4 h. Cortex showed mean 24, 46, 78 damaged neurons after 5-HTP 50, 75 and 150 mg doses. Thalamus and hypothalamus showed moderate damage in nerve cells and medulla and brainstem showed significant but mild neuronal damages (Table 5). These damaged neurons are present in the edematous brain areas exhibiting BBB breakdown.

4.8 Effect of 5-HTP on lanthanum exudation across the endothelial cells

Ultrastructural studies showed lanthanum extravasation across the endothelial cells in 5-HTP treated rats after 4 h. The magnitude and intensity of lanthanum exudation within the endothelial cell cytoplasm and in basal lamina occasionally is dose dependent (results not shown). Lanthanum could be sometimes seen within the microvesicular profiles in the endothelial cell cytoplasm as well as in the basal lamina while tight junctions are closed to this tracer in all the microvessels examined (Fig. 2). Interestingly, one vessel could show lanthanum exudation in some part of the endothelial cell while the adjacent endothelial cell is completely free of lanthanum exudation within the cell cytoplasm or across the endothelial cell. In this regard, the vessel diameter or number of endothelial cells affected by lanthanum exudation in one capillary is not consistent. However, those capillaries showing exudation of lanthanum across the cell cytoplasm and in basal lamina perivascular edema id often seen in 5-HTP treated rats. The number of microvascular showing leakage of lanthanum following 5-HTP treatments is related to the doses of the compound used. However, the pattern of lanthanum exudation and mode of tracer transfer is not altered by the dose used. Tight junctions in all cases and in all doses of 5-HTP remained intact to lanthanum (results not shown).

Normal saline treated animals showed lanthanum as black particles totally confined within the lumen of the microvessels.

4.9 Effects of 5-HTP on albumin and neuronal nitric oxide synthase immunohistochemistry

5-HTP treatment exhibited albumin leakage in the neuropil of cerebral cortex, cerebellum, hippocampus, thalamus and hypothalamus in the areas showing edematous swelling and sponginess. The magnitude and intensity of albumin leakage is dose dependent in 5-HTP treated groups. Medulla and brainstem showed fewer but significantly higher number of albumin positive cells as compared to control group or saline treated animals (Figs. 4 and 5).

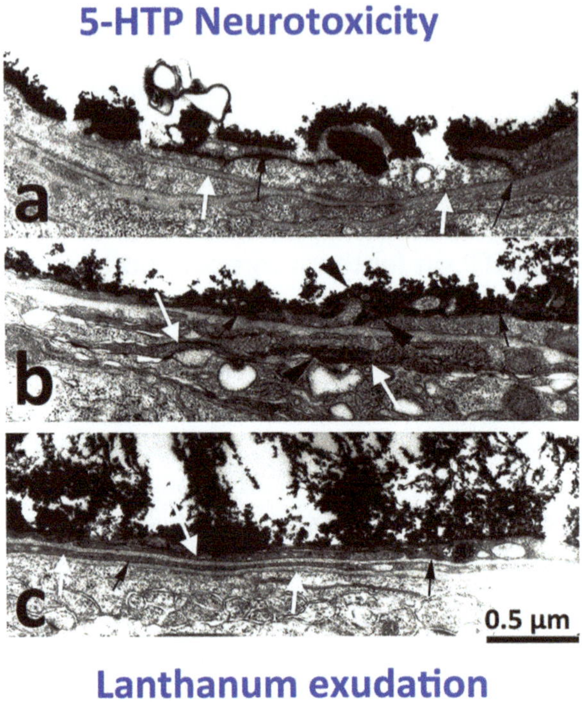

Fig. 4 High power transmission electron micrograph (TEM) showing lanthanum exudation across the cerebral endothelial cells from parietal cerebral cortex of one 5-HTP (150 mg/kg, i.p.) administered rats after 4 h (a–c). Lanthanum is largely confined within the lumen of the cerebral endothelium and stopped at the tight junctions (black arrow, a) and presence of lanthanum in the basal lamina (white arrow) is not seen (a). Lanthanum is seen into the basal lamina (white arrow, b) whereas; the tight junction (black arrow) is still impermeable to the electron-dense tracer (b). Lanthanum infiltration into the basal lamina of one capillary (black arrows) whereas the tight junctions are still intact (white arrow, c). Bar = 500 nm.

Interestingly, upregulation of nNOS is also seen in the vicinity of albumin exudation in the neuropil. This upregulation of nNOS is most prominent in the cerebellum followed by hippocampus and cerebral cortex. Thalamus and hypothalamus also showed significant nNOS upregulation in the edematous region. Very few nNOS positive ells are seen in the medulla and brainstem. The nNOS immunoreactivity is most prominent in the neuronal cell cytoplasm and occasionally karyoplasm also showed nNOS immunoreaction in 5HTP treated group. The magnitude and intensity of nNOS upregulation showed a close parallelism with the dose of the compound used.

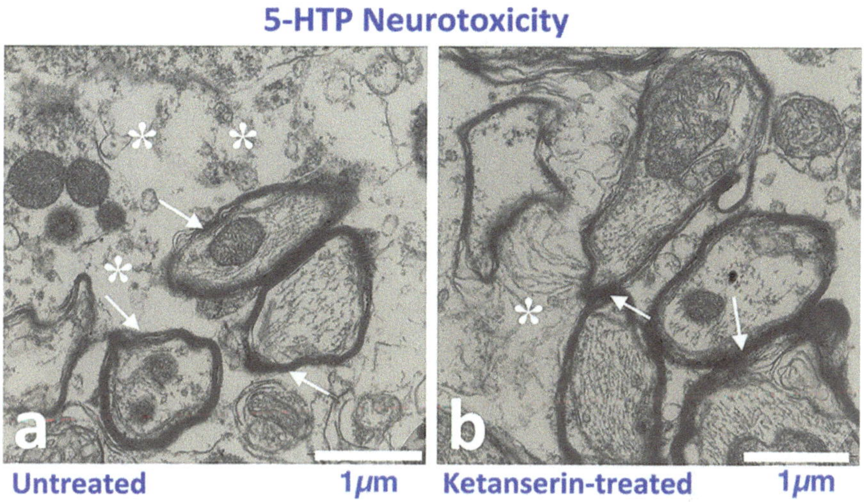

Fig. 5 High power transmission electron micrograph of neuropil from IIIrd cellular layer of parietal cerebral cortex in one 5-HTO (150mg/kg, i.p.) administered rats after 4h (a) and neuroprotection by ketanserin (b). Vacuolation in neuropil, edema (*) and axonal distortion is clearly seen in untreated 5-HTP administered rats (a). In ketanserin pretreated rat 5-HTP administration induced changes in the neuropil is markedly less evident (*). The neuropil appears to be compact (b) than untreated (a) group after 5-HTP administrations. Bar = 1 μm.

5. Effect of drug treatments on 5-HTP induced alteration in behavioral, physiological and cerebrovascular function

Since, 150mg 5-HTP produced the most increase in the 5-HT levels in the plasma, BBB permeability, CBF alterations, stress symptoms, physiological variables, the influence of several drugs on these parameters were observed in this group only. The following drugs were used.

(a) *Effect of tryptophan hydroxylase inhibitor, p-CPA*

Pretreatment with *p*-CPA significantly reduced the BBB permeability increase to Evans blue and radioactive iodine (Table 6). The regional BBB permeability was also significantly attenuated. The regional CBF was almost restored near normal levels. This treatment significantly thwarted the 5-HT increase in the brain and spinal cord (Table 7).

There was a significant reduction in core body temperature from the untreated group. The 5-HTP induced tachycardia was mainly absent. Microhemorrhages in the stomach wall and diarrheic stool were significantly reduced by *p*-CPA treatment. The other behavioral symptoms were considerably reduced (Table 6).

Table 6 Effect of 5-HTP (175 mg/kg) on regional blood-brain barrier permeability and regional cerebral blood flow.

	rBBB permeability [131]Iodine %			rCBF mL/g/min		
Brain regions	**Saline**	**5-HTP**	**% Change**	**Saline**	**5-HTP**	**% Change**
1. Cingulate cortex	0.18 ± 0.03	1.48 ± 0.08**	+722	1.04 ± 0.04	0.82 ± 0.06**	−21
2. Frontal cortex	0.24 ± 0.06	0.66 ± 0.10*	+175	1.16 ± 0.06	0.80 ± 0.04**	−31
3. Parietal cortex	0.30 ± 0.04	1.76 ± 0.08**	+486	1.23 ± 0.04	0.84 ± 0.08**	−32
4. Occipital cortex	0.32 ± 0.04	1.46 ± 0.11***	+356	1.18 ± 0.06	0.85 ± 0.06***	−28
5. Temporal cortex	0.28 ± 0.05	1.64 ± 0.12***	+485	1.09 ± 0.10	0.82 ± 0.10*	−25
6. Hippocampus	0.26 ± 0.08	1.76 ± 0.08***	+576	0.96 ± 0.04	0.78 ± 0.08**	−20
7. Caudate nucleus	0.28 ± 0.04	1.64 ± 0.06***	+485	0.94 ± 0.06	0.72 ± 0.08**	−23
8. Cerebellar vermis	0.18 ± 0.02	1.58 ± 0.04***	+777	0.98 ± 0.05	0.74 ± 0.05**	−24
9. Cerebellar cortex	0.16 ± 0.04	1.64 ± 0.08***	+925	0.92 ± 0.04	0.72 ± 0.02***	−21
10. Colliculus superior	0.25 ± 0.05	1.45 ± 0.10***	+480	1.21 ± 0.08	0.83 ± 0.10*	−31
11. Colliculus inferior	0.20 ± 0.07	0.34 ± 0.08	+7	1.05 ± 0.06	0.96 ± 0.08	−8.5
12. Thalamus	0.44 ± 0.05	1.68 ± 0.08**	+281	0.87 ± 0.04	0.76 ± 0.06*	−13
13. Hypothalamus	0.48 ± 0.08	1.89 ± 0.06**	+293	0.92 ± 0.06	0.80 ± 0.08***	−13
14. Pons	0.23 ± 0.04	0.78 ± 0.08***	+239	1.01 ± 0.03	0.86 ± 0.08**	−15
15. Medulla	0.18 ± 0.06	0.56 ± 0.05**	+211	1.14 ± 0.08	0.94 ± 0.06*	−17.5
16. Brain stem	0.19 ± 0.04	1.58 ± 0.10*	+731	1.18 ± 0.04	1.04 ± 0.10	−12
Average	0.28 ± 0.09	1.37 ± 0.49	+451	1.05 ± 0.11	0.83 ± 0.08	−22
		+389			−21	

Values are mean ± SD, * $P < 0.05$; ** $P < 0.01$; ***$P < 0.001$, compared from control group; Student's unpaired t-test.
Figures below the values represent % change from control group. + = increase, − = decrease.

(b) *Effect of cyclooxygenase inhibitor, indomethacin*

Pretreatment with indomethacin significantly reduced the stress symptoms and physiological variables. However, indomethacin did not reduce the rise in rectal temperature caused by 5-HTP administrations. Increase in heart rate and respiration, occurrence of gastric hemorrhages, diarrheic stool, wet dog shakes, jumping behavior, motor behavior were considerably attenuated by indomethacin treatment, Blood gases and blood pressure were not statistically different from 5-HTP-treated group.

The BBB permeability was significantly reduced in indomethacin treated animals. The CBF decline in this group was absent.

There was a slight but significant decrease in 5-HT content in this drug treated group. Thus, hippocampus, cerebellum and thalamus showed only mild increase in 5-HT (18–139%). Hypothalamus did not show any change in 5-HT content. The plasma 5-HT level showed only 50% increase compared to untreated group.

(c) *Effect of anxiolytic drug, diazepam*

Pretreatment with diazepam significantly inhibited stress and behavioral symptoms without affecting 5-HTP induced rise in the body temperature (Table 6). This drug markedly inhibited 5-HTP induced gastric hemorrhages. Alterations in motor behavioral pattern were completely absent in diazepam treated rats. Tachycardia was absent in these animals without affecting respiration. The MABP continued to remain depressed like untreated rats and no significant difference in the physiological variables were seen compared to 5-HTP administered rats.

Diazepam pretreated rats did not exhibit 5-HTP induced breakdown of the BBB permeability to Evans blue or radioactive iodine. The CBF values were also restored to normal levels. Regional BBB permeability showed significant increase in radiotracer in two out of 16 brain regions (medulla and inferior colliculus) from the control group. The decline in regional CBF was mild (ranging from +1% to −17%). In one out of 16 regions, the decline in regional CBF was statistically significant (occipital cortex).

(d) *Effect of a potent 5-HT2 and a weak Histamine H1 receptor antagonist, cyproheptadine*

Cyproheptadine pretreatment influenced stress and behavioral symptoms mildly. Thus, behavioral symptoms and hyper motor activity were less evident in this drug treated rats following n 5-HTP administration. Mild diarrheic stool was noted in this group. The incidence of wet dog shakes was considerably reduced. This drug treatment prevented tachycardia and respiratory changes caused by 5-HTP. However, 5-HTP induced hypotension and changes in blood gases or arterial pH were not affected significantly (Table 6).

Cyproheptadine pretreatment significantly reduced the extravasation of Evans blue and radioactive iodine in the brain following 5-HTP injections. The CBF also did not exhibit significant decline from the untreated control group. Changes in the regional BBB and CBF were also markedly influenced by cyproheptadine pretreatment. In this drug treated group 5-HTP administration was able to retain significant increase in the permeability of the radiotracer in medulla only. In all 15 other brain regions, the mild increase in the BBB permeability by 9–125% was not found statistically significant. The CBF showed significant decline in parietal cortex and in the superior colliculus. In other 14 regions, changes in rCBF (+1% to −17%) were not significant.

Cyproheptadine pretreatment in general did not influence regional 5-HT increase following 5-HTP administrations. In cerebellum and in the spinal cord, a significantly higher serotonin content was noted in this drug treated animals compared to 5-HTP treatments alone (Table 7).

(e) *Effect of a potent and specific 5-HT2-receptor antagonist, ketanserin*

Pretreatment with ketanserin significantly attenuated most of the stress and behavioral symptoms without affecting 5-HTP induce rise in the core body temperature. Tachycardia and hyperpnoea were absent in this drug treated rats. However, state of 5-HTP induced hypotension and changes in the blood gases or arterial pH were not significantly different (Table 6). Behavioral symptoms and hyper motor activity were less marked in this drug treated animals following 5-HTP challenges. Gastric hemorrhages were significantly reduced without any significant effect on diarrhea or soft stool.

The BBB permeability to Evans blue and radiotracer were markedly attenuated. The CBF decline was much less evident and insignificant compared to control group. The region changes in the BBB and CBF were also less intense. Only spinal cord exhibited 89% increase in iodine extravasation compared to control group. In all other 15-brain regions the increase in radiotracer extravasation from 0% to 100% were not found statistically significant. The regional CBF changes ranged from +2% to −12% from the control group and were insignificant.

Changes in regional serotonin content were not significantly affected by ketanserin in this group caused by 5-HTP injections. In spinal cord, 5-HT content was significantly much higher in ketanserin treated group compared to untreated rats (Table 6).

Table 7 Pharmacological manipulation of 5-HTP (150 mg/kg) induced changes in the rat brain.

Prototype	Known function	Dose/route (mg/kg)	Schedule	Source
1. 5-Hydroxytryptophan	Serotonin precursor	50, 75, 150, i.p.	4 h before	Sigma Chemicals, USA
2. *p*-Chlorophenylalanine	Serotonin synthesis inhibitor	100/day for 3 days, i.p.	1 day before	Sigma Chemicals, USA
3. Indomethacin[b]	Prostaglandin synthesis inhibitor	10, i.p.	30 min before	Sigma Chemicals, USA
4. Diazepam (Valium)[a]	Anxiolytic drug	4, s.c.	30 min before	ICI, UK
5. Cyproheptadine	5-HT2 receptor antagonist	5, i.p.	30 min before	Merck, Sharp & Dhome, UK
6. Ketanserin	Selective 5-HT2 receptor antagonist	1, i.p.	30 min before	Lysen, Beerse, Belgium
7. Vinblastine[a]	Antimitotic and vesicular transport inhibitor	0.8, i.v.	30 min before	Eli, Lilly & Co., USA

[a]Ampoule.
[b]Dissolved in water at 60 °C with a pinch of sodium bicarbonate and pH adjusted to 7 before administration.
The choice of drug-dosage, route and time schedule were determined according to the standard protocol from literature (for details see text); Other drugs in powder form were dissolved in sterile water and injected i.p. (0.3 mL), i.v. (0.5 mL), s.c. (0.05 mL).

(f) *Effect of a vesicular transport inhibitor and antimitotic drug, vinblastine*

Pretreatment with vinblastine in general did not influence 5-HTP induced stress and behavioral symptoms markedly. The body temperature continues to remain high and the tachycardia and hyperpnoea were present. Diarrheic stool with hyper locomotion and other motor behavior was apparent. The arterial pH and blood gases were very similar to that of untreated 5-HTP injected rats.

The extraction of dye and radiotracer were, however, significantly reduced from the untreated group. The absolute extravasations of tracers were significantly different from control group. However, the CBF continued to decline like untreated 5-HTP challenged group (Table 7). The regional changes in the BBB were significantly less than the untreated 5-HTP administered groups except in medulla, where the extravasation of radiotracer was 55% higher than the control group. Changes in BBB permeability in other regions ranged from −14% to 137% from the control, but the differences were not found statistically significant. The regional decline in CBF was significant (−17% to −28%) in 10 out of 16 regions. In remaining six regions a decline in 2–17% was not significant from control group.

The serotonin content was very similar to that of untreated 5-HTP administered group. In hippocampus, the 5-HT content was significantly higher in vinblastine treated group compared to untreated 5-HTP administered rats.

6. Effects of drugs on 5-HTP induced brain pathology and immunohistochemistry

Morphological and ultrastructural studies were done in 150 mg dose of 5-HTP treated animals only and its modulation by few-selected drug such as *p*-CPA, diazepam and ketanserin.

Our observations showed that pretreatment with these drug significantly reduced neuronal changes in the cerebral cortex, hippocampus and cerebellum caused by 5-HTP administrations. Also albumin immunoreactivity and nNOS upregulation was significantly reduced in these brain areas in drug treated group after 5-HTP administrations. The most superior neuroprotective effects on cell changes and albumin or nNOS immunoreaction in 5-HTP treated groups was induced by diazepam followed by *p*-CPA and ketanserin (results not shown).

Ultrastructural changes on lanthanum exudation as also reduced by these drugs after 5-HTP administration. Thus, lanthanum was largely confined within the lumen in these drug treated animals after 5-HTP administrations. However, only a few endothelial cells in sporadic capillaries exhibited vesicular transport of lanthanum within the endothelial cell cytoplasm. Present of lanthanum in the basal lamina is not seen in these drug treated animals after 5-HTP administrations.

7. Possible mechanisms of 5-HTP induced neurotoxicity

The salient new findings of the present investigation show that acute 5-HTP administration is capable of inducing marked breakdown of the BBB permeability. This observation is the first to point out that this breakdown of the BBB could be one of the most important factors in causing disturbed mental function and mood alterations on the first day of the drug therapy (Das, Bagchi, Bagchi, & Preuss, 2004). Altered BBB function to serum factors is the hallmark of several psychiatric diseases (Montagne et al., 2016, 2017; Ozkizilcik et al., 2018; Sharma et al., 2012; Sweeney, Montagne, et al., 2019). However, it is still not known whether this increased permeability of the BBB seen in depression, mental retardation, schizophrenia, and autism will adversely affect 5-HTP administration induced BBB breakdown as compared to normal healthy brain (see Birdsall, 1998). This is an important novel subject that requires additional investigation.

Fortunately, this increase in BBB permeability to proteins is time-dependent and appears to be reversible in nature. This is evident from the finding that no further evidence of extravasation of Evans blue or radioactive iodine was apparent when the BBB was examined 12h after 5-HTP administration. However, it is unclear whether the BBB is still open to small molecules as compared to serum protein complex (MW 28–68kDa) (Bradbury, 1979; Rapoport, 1976). There are evidences that hyperosmotic solutions induced BBB breakdown to Evans blue is repaired within 2h period whereas, the BBB breakdown to Na+ is still open for 4 days after the initial insult (see Rapoport, 1976). This suggests that 5-HTP induced breakdown of the BBB could have long lasting effects leading to potentially long-term brain dysfunction.

The altered BBB permeability appears to be related with its ability to enhance endogenous serotonin concentrations in the plasma and brain (Udenfriend et al., 1957). Increased circulating serotonin level markedly affects CBF (Bonvento et al., 1994; Burnstock, 1985; Edvinsson & MacKenzie, 1976; Kobayashi et al., 1985; MacKenzie & Scatton, 1987; Raskin, 1981). This idea is further supported by the fact that 5-HTP-administration resulted in widespread alterations in the regional cerebral blood flow (rCBF) in present investigation. Although, elevation of endogenous serotonin levels following 5-HTP administrations has been known for decades, to our knowledge, this is the first report that demonstrates widespread alterations in the BBB permeability and the CBF in several brain regions after its systemic administration.

To examine breakdown of the BBB permeability, we used two different protein tracers (Sharma, 2007; Sharma et al., 1990a; Sharma & Dey, 1981; Sharma, Dey, & Olsson, 1989; Sharma, Olsson, & Dey, 1989, 1995; Sharma, Westman, Cervós-Navarro, et al., 1995; Winkler et al., 1995). Both the tracers bind to endogenous serum proteins when administered in the systemic circulation (Bradbury, 1979; Rapoport, 1976; Sharma, 1999, 2009;

Sharma & Westman, 2003, 2004). Thus, leakage of these tracers in brain in fact represent extravasation of tracer-protein complex (Bradbury, 1979; Brightman et al., 1970; Rapoport, 1976; Sharma et al., 1990a). One molecule of Evans blue (molecular weight 960), in particular, will bind to approximately 12 molecules of albumin (combined molecular wt. 68,000) (Bradbury, 1979; Rapoport, 1976; Sharma, 1982). About two-thirds of serum albumin will be bound to Evans blue in the doses that is used in this study (Sharma, 2007; Sharma et al., 1990a; Sharma & Dey, 1981; Sharma, Dey, & Olsson, 1989; Sharma, Olsson, & Dey, 1989, 1995; Sharma, Westman, Cervós-Navarro, et al., 1995; Winkler et al., 1995). Likewise, iodine largely binds to albumin by 50% (mol. wt. about 28,000) (Bradbury, 1979; Rapoport, 1976). Thus, the wide permeation of iodine compared to Evans blue may be due to differences in the molecular sizes and diameter of the tracer-protein or complexes and/or due to the differences in the proteins to which they bind in circulation (Sharma, 2007; Sharma et al., 1990a; Sharma & Dey, 1981; Sharma, Dey, & Olsson, 1989; Sharma, Olsson, & Dey, 1989, 1995; Sharma, Westman, Cervós-Navarro, et al., 1995; Winkler et al., 1995). Obviously, smaller molecular size tracers could pass through the barrier more readily compared to larger sized tracers at the BBB interface.

Leakage of plasma proteins into the cerebral microenvironment leads to vasogenic edema (Sharma et al., 1991a, 1992, 1994). This is clearly evident with our findings by 5-HTP induced brain edema formation. Brain edema and volume swelling is present in almost all brain regions exhibiting BBB breakdown to proteins. Leakage of endogenous albumin using immunohistochemical techniques further confirms this hypothesis in present investigation. Thus, albumin leakage is evident in areas showing sponginess and edematous expansion. Furthermore, distorted neurons, swollen or shrunken with perineuronal edema is present in areas showing albumin immunoreaction in the neuropil after 5-HTP administrations. When the BBB breakdown of Evans blue and radioiodine is no longer observed after 12 h of 5-HTP administrations, neuronal damages, leakage of serum albumin is till visible within the neuropil. These observations clearly suggests that 5-HTP induced BBB dysfunction leads to long-term neuronal damages and may lead to neurodegeneration over time (H.S. Sharma, unpublished observations).

The possibility that 5-HTP induced BBB breakdown, edema formation and cell injury could also be associated with increased oxidative stress and free radical formation (Kayacan, Yazar, Cerit, & Ghojebeigloo, 2019; Reyes-Gonzales et al., 2009). This idea is supported by the fact that we have observed upregulation of neuronal nitric oxide synthase (nNOS) in various parts of the brain associated with neuronal injuries (Chae et al., 2009; Henry, Ishimura, & Peisach, 1976). The nNOS upregulation suggests generation of nitric oxide $(NO)^-$ the free radical that induces membrane damage directly (Dalkara & Moskowitz, 1994; Dawson, Dawson, & Snyder, 1992; Dawson, Zhang, Dawson, & Snyder, 1994;Sharma et al., 1996; Sharma et al., 2005; Sharma, Alm, & Westman, 1998; Sharma, Nyberg, et al., 1998). This damage to endothelial cells leads to BBB breakdown (Sharma et al., 1996; Sharma et al., 2005; Sharma, Alm, & Westman, 1998; Sharma, Nyberg, et al., 1998). Massive upregulation of nNOS by 5-HTP in a dose dependent manner suggest that 5-HTP-induced neurotoxicity could be mediated through NO production (see Sharma, Alm, & Westman, 1998). High level of circulating serotonin could also one of the primary reasons to induce oxidative stress and production of free radical NO (Bedrosian & Nelson, 2014; Dhir & Kulkarni, 2011; Lee et al., 2013; Zhou, Zhu, Nemes, & Zhu, 2018). This fact is well reflected in upregulation of nNOS immunoreactivity in 5-HTP administered group.

That 5-HTP induced nNOS upregulation could induce endothelial cell membrane disruption leading to BBB breakdown is further supported by our ultrastructural investigation of microvascular permeability using the electron dense tracer lanthanum (12 Å diameter). 5-HTP administrations in a dose dependent manner induce vesicular transport of lanthanum across the endothelial cell in several places in the brain exhibiting nNOS upregulation. Lanthanum is also seen in the basal lamina. However, the tight junctions remain intact. This suggests that serotonin, NO and oxidative stress could somehow stimulate vesicular transport at the endothelial cells leading to the breakdown of the BBB following 5-HTP administrations (Sharma, 2007; Sharma et al., 1990a, 1990b, 1991a, 1992, 1994; Sharma & Dey, 1981, 1987, 1988; Sharma, Dey, & Olsson, 1989; Sharma, Olsson, & Dey, 1989, 1995; Sharma, Westman, Cervós-Navarro, et al., 1995; Winkler et al., 1995).

The mechanisms of BBB breakdown following 5-HTP administrations appear to be related with several other factors stress associated with 5-HTP administrations. Stress is one of the main reasons to enhance plasma and brain serotonin level, hyperthermia and free radical formation (Dey & Sharma, 1983, 1984; Sharma et al., 1990b; Sharma & Dey, 1984, 1986a, 1986b, 1987, 1988).

Administration of 5-HTP also resulted in marked stress symptoms and behavioral alteration. These changes are most prominent at the time of increased BBB permeability. However, it is unclear whether altered behavioral and symptoms are due to effects of the drug alone or drug-induced elevation of 5-HT level. Since several drugs in this study influence symptoms and behavioral alterations, it seems most likely that behavioral symptoms are related with both the peripheral and central actions of the 5-HT (Musumeci et al., 2015).

However, prevention of BBB permeability by several drugs in this investigation, irrespective of their effects on 5-HT level, modified the behavioral alterations and stress symptoms. Thus, it seems quite likely that breakdown of the BBB is somehow interrelated with behavioral alterations and/or stress symptoms (Dey & Sharma, 1983, 1984; Sharma, 2007; Sharma et al., 1990a, 1990b, 1991a, 1992, 1994; Sharma & Dey, 1981, 1984, 1986a, 1987, 1988; Sharma, Dey, & Olsson, 1989; Sharma, Olsson, & Dey, 1989, 1995; Sharma, Westman, Cervós-Navarro, et al., 1995; Winkler et al., 1995). Alternatively, symptoms associated with 5-HTP administrations such as hyperthermia and stress caused by elevation of serotonin may also contribute to BBB permeability (Dey & Sharma, 1983, 1984; Sharma et al., 1990b, 1991a, 1992, 1994; Sharma & Dey, 1984, 1986a, 1987, 1988). Obviously, serotonin induced stress response can be modulated by drugs modulating serotonergic transmission and/or its receptor function.

5-HTP administration results in profound alterations in the rCBF. Although, those brain areas showing regional BBB (rBBB) breakdown also exhibit reduction in the rCBF, however, rCBF is not directly responsible for rBBB breakdown. This is evident with the fact that the magnitude of reduction in rCBF is unrelated with the intensity of the rBBB breakdown.

This is further apparent from the fact that drugs that reduced serotonin levels in plasma and brain prevented BBB breakdown, edema formation, nNOS expression and cell injuries. Whereas, drugs that are receptor blocker or inhibitor of vesicular transport did not prevent serotonin elevation after 5-HTP administrations but blocked the BBB breakdown, edema formation cell injury and albumin extravasation (H.S. Sharma, unpublished observation). Obviously, blockade of serotonin receptors could not stimulate cascades of molecular signaling leading to BBB breakdown (Olsson et al., 1992; Sharma, 2007; Sharma et al., 1990a, 1990b; Sharma, Dey, & Olsson, 1989; Sharma, Olsson, & Dey, 1989, 1995; Sharma, Westman, Cervós-Navarro, et al., 1995; Winkler et al., 1995). Inhibition of microtubule function or vesicular transport can also prevent BBB leakage because the 5-HTP induced BBB breakdown is mediated through enhanced vesicular transport and not by widening of the tight junctions (Sharma, 2007; Sharma et al., 1990a; Sharma & Dey, 1981; Sharma, Dey, & Olsson, 1989;

Sharma, Olsson, & Dey, 1989, 1995; Sharma, Westman, Cervós-Navarro, et al., 1995; Winkler et al., 1995). Obviously, when BBB breakdown is prevented it will reduce edema formation and subsequently nNOS upregulation and cellular injury.

8. Conclusions and future perspectives

In conclusion, our observations suggest that acute administration of 5-HTP induces breakdown of the BBB to serum proteins leading to edema formation and neuronal injuries. These changes are associated with enhanced plasma and brain serotonin level associated with nNOS upregulation. Leakage of proteins although reduced over 12h but the deposition of plasma protein and cell injury could continue for longer periods after 6-HTP administrations. Further research is needed to understand 5-HTP induced acute and chronic neurotoxicity. Since oxidative stress appears to play key roles in 5-HTP neurotoxicity adjunct therapies using neurotrophic factors and related strategies are needed to contain acute brain damage. Our laboratory is engaged in finding suitable adjunct therapy to minimize 5-HTP neurotoxicity by coadministration of cerebrolysin-a balanced composition of several neurotrophic factors and active peptide fragment either alone or its nanodelivery. These strategies could expand our knowledge and help cases in psychiatric diseases where 5.HTP therapy is needed.

Acknowledgments

This investigation was supported by Grants from Swedish Medical Research Council Nr. 2710; Grants from the Alzheimer's Association (IIRG-09-132087), the National Institutes of Health (R01 AG028679) and the Dr. Robert M. Kohrman Memorial Fund (M.A.S., R.J.C.); Society for Neuroprotection and Neuroplasticity (SSNN), Romania; Astra Zeneca, Mölndal; Göran Gustafsson Foundation, Stockholm, Sweden; Alexander von Humboldt Foundation, Bonn, Germany; The University Grants Commission, New Delhi, India; Indian Council of Medical Research, New Delhi, India. The skillful technical assistance of Mr. Om Prakash Gupta, Shiv Mandir Singh (Varanasi, India); Katja Deparade, Franzisca Drum, Elisabeth Scherer (Berlin, Germany); Secretarial assistance of Angela Ludwig (Berlin, Germany), Madeline Järild, Gunilla Åberg and Gunilla Tibling (Uppsala, Sweden) were highly appreciated.

References

Andén, N. E., & Modigh, K. (1972). Effects of p-chlorophenylalanine and a monoamine oxidase inhibitor on the 5-hydroxytryptamine in the spinal cord and transection. *Journal of Neural Transmission*, *33*(3), 211–222 [No abstract available].

Anon. (1989). Drugs acting on 5-hydroxytryptamine receptors. *Lancet*, *2*(8665), 717–719 [Review. No abstract available].

Bedrosian, T. A., & Nelson, R. J. (2014). Nitric oxide and serotonin interactions in aggression. *Current Topics in Behavioral Neurosciences*, *17*, 131–142. https://doi.org/10.1007/7854_2013_273.

Bhagavan, H. N., & Coursin, D. B. (1973). Effects of pyridoxine deficiency and DL-p-chlorophenylalanine administration to rats on 5-hydroxytryptamine and noradrenaline concentrations in brain and 5-hydroxytryptamine concentration in blood. *The Biochemical Journal*, *134*(3), 763–767.

Birdsall, T. C. (1998). 5-Hydroxytryptophan: A clinically-effective serotonin precursor. *Alternative Medicine Review*, *3*(4), 271–280.

Bonvento, G., MacKenzie, E. T., Seylaz, J., & Lacombe, P. (1994). The cerebrovascular role of the ascending serotonergic system: New vistas. *Journal of the Autonomic Nervous System*, *49*(Suppl), S37–S42 [Review. No abstract available].

Bosma, E. K., van Noorden, C. J. F., Schlingemann, R. O., & Klaassen, I. (2018). The role of plasmalemma vesicle-associated protein in pathological breakdown of blood-brain and blood-retinal barriers: Potential novel therapeutic target for cerebral edema and diabetic macular edema. *Fluids Barriers CNS*, *15*(1), 24. https://doi.org/10.1186/s12987-018-0109-2 [Review].

Bradbury, M. W. B. (1979). *The concept of a blood-brain barrier*. Chichester, London: Wiley.

Brightman, M. W., Klatzo, I., Olsson, Y., & Reese, T. S. (1970). The blood-brain barrier to proteins under normal and pathological conditions. *Journal of the Neurological Sciences*, *10*(3), 215–239.

Burnstock, G. (1985). Neurogenic control of cerebral circulation. *Cephalalgia*, *5*(Suppl. 2), 25–33 [Review].

Chae, H. S., Kang, O. H., Choi, J. G., Oh, Y. C., Lee, Y. S., Jang, H. J., et al. (2009). 5-hydroxytryptophan acts on the mitogen-activated protein kinase extracellular-signal regulated protein kinase pathway to modulate cyclooxygenase-2 and inducible nitric oxide synthase expression in RAW 264.7 cells. *Biological & Pharmaceutical Bulletin*, *32*(4), 553–557.

Chakraborty, S., Lennon, J. C., Malkaram, S. A., Zeng, Y., Fisher, D. W., & Dong, H. (2019). Serotonergic system, cognition, and BPSD in Alzheimer's disease. *Neuroscience Letters*, *704*, 36–44. https://doi.org/10.1016/j.neulet.2019.03.050 [Epub ahead of print. Review].

de Clerck, F., David, J. L., & Janssen, P. A. (1982). Inhibition of 5-hydroxytryptamine-induced and -amplified human platelet aggregation by ketanserin (R 41 468), a selective 5-HT2-receptor antagonist. *Agents and Actions*, *12*(3), 388–397.

Conio, B., Martino, M., Magioncalda, P., Escelsior, A., Inglese, M., Amore, M., et al. (2019). Opposite effects of dopamine and serotonin on resting-state networks: Review and implications for psychiatric disorders. *Molecular Psychiatry*. https://doi.org/10.1038/s41380-019-0406-4 [Epub ahead of print. Review].

Crook, D., & Collins, A. J. (1977). Comparison of effects of aspirin and indomethacin on human platelet prostaglandin synthetase. *Annals of the Rheumatic Diseases*, *36*(5), 459–463.

Dailly, E., Hascoët, M., Colombel, M. C., Jolliet, P., & Bourin, M. (2002). Relationship between cerebral pharmacokinetics and anxiolytic activity of diazepam and its active metabolites after a single intra-peritoneal administration of diazepam in mice. *Human Psychopharmacology*, *17*(5), 239–245.

Dalkara, T., & Moskowitz, M. A. (1994). The complex role of nitric oxide in the pathophysiology of focal cerebral ischemia. *Brain Pathology*, *4*(1), 49–57 [Review].

Das, Y. T., Bagchi, M., Bagchi, D., & Preuss, H. G. (2004). Safety of 5-hydroxy-L-tryptophan. *Toxicology Letters*, *150*(1), 111–122 [Review].

Daut, R. A., & Fonken, L. K. (2019). Circadian regulation of depression: A role for serotonin. *Frontiers in Neuroendocrinology*. https://doi.org/10.1016/j.yfrne.2019.04.003 [Epub ahead of print. Review].

Dawson, T. M., Dawson, V. L., & Snyder, S. H. (1992). A novel neuronal messenger molecule in brain: The free radical, nitric oxide. *Annals of Neurology*, *32*(3), 297–311 [Review].
Dawson, T. M., Zhang, J., Dawson, V. L., & Snyder, S. H. (1994). Nitric oxide: Cellular regulation and neuronal injury. *Progress in Brain Research*, *103*, 365–369 [Review. No abstract available].
Dey, P. K., & Sharma, H. S. (1983). Ambient temperature and development of traumatic brain oedema in anaesthetized animals. *The Indian Journal of Medical Research*, *77*, 554–563 [No abstract available].
Dey, P. K., & Sharma, H. S. (1984). Influence of ambient temperature and drug treatments on brain oedema induced by impact injury on skull in rats. *Indian Journal of Physiology and Pharmacology*, *28*(3), 177–186.
Dhir, A., & Kulkarni, S. K. (2011). Nitric oxide and major depression. *Nitric Oxide*, *24*(3), 125–131. https://doi.org/10.1016/j.niox.2011.02.002 [Epub 2011 Feb 16].
Doggrell, S. A. (1992). Further analysis of the inhibitory effects of dihydroergotamine, cyproheptadine and ketanserin on the responses of the rat aorta to 5-hydroxytryptamine. *Journal of Autonomic Pharmacology*, *12*(4), 223–236.
Dustin, P., Jr. (1963). New aspects of the pharmacology of antimitotic agents. *Pharmacological Reviews*, *15*, 449–480 [Review. No abstract available].
Edvinsson, L., & MacKenzie, E. T. (1976). Amine mechanisms in the cerebral circulation. *Pharmacological Reviews*, *28*(4), 275–348 [Review. No abstract available].
Elliott, K. A., & Jasper, H. (1949). Measurement of experimentally induced brain swelling and shrinkage. *The American Journal of Physiology*, *157*(1), 122–129 [No abstract available].
Erickson, M. A., & Banks, W. A. (2018). Neuroimmune axes of the blood-brain barriers and blood-brain interfaces: Bases for physiological regulation, disease states, and pharmacological interventions. *Pharmacological Reviews*, *70*(2), 278–314. https://doi.org/10.1124/pr.117.014647 [Review].
Fahn, S. (1977). New approaches in the management of hyperkinetic movement disorders. *Advances in Experimental Medicine and Biology*, *90*, 157–173.
Florian, S., & Mitchison, T. J. (2016). Anti-microtubule drugs. *Methods in Molecular Biology*, *1413*, 403–421. https://doi.org/10.1007/978-1-4939-3542-0_25.
Fukuda, K. (2014). 5-HTP hypothesis of schizophrenia. *Medical Hypotheses*, *82*(1), 20–23. https://doi.org/10.1016/j.mehy.2013.10.026 [Epub 2013 Nov 8].
Fukuda, K. (2015). Integrated theory to unify status among schizophrenia and manic depressive illness. *Medical Hypotheses*, *85*(4), 506–511. https://doi.org/10.1016/j.mehy.2015.06.012 [Epub 2015 Jun 24].
Gafni, Y., Schwartzman, M., & Raz, A. (1978). Prostaglandin biosynthesis in rabbit kidney medulla: Inhibition in-vitro vs. in-vivo by aspirin, indomethacin and meclofenamic acid. *Prostaglandins*, *15*(5), 759–772.
Gordh, T., Chu, H., & Sharma, H. S. (2006). Spinal nerve lesion alters blood-spinal cord barrier function and activates astrocytes in the rat. *Pain*, *124*(1–2), 211–221 [Epub 2006 Jun 27].
Haberzettl, R., Bert, B., Fink, H., & Fox, M. A. (2013). Animal models of the serotonin syndrome: A systematic review. *Behavioural Brain Research*, *256*, 328–345. https://doi.org/10.1016/j.bbr.2013.08.045 [Epub 2013 Sep 1. Review].
Heistad, D. D., Marcus, M. L., & Mueller, S. (1977). Measurement of cerebral blood flow with microspheres. *Archives of Neurology*, *34*(11), 657–659.
Henry, Y., Ishimura, Y., & Peisach, J. (1976). Binding of nitric oxide to reduced L-tryptophan-2,3-dioxygenase as studied by electron paramagnetic resonance. *The Journal of Biological Chemistry*, *251*(6), 1578–1581.
Hurtado-Alvarado, G., Domínguez-Salazar, E., Pavon, L., Velázquez-Moctezuma, J., & Gómez-González, B. (2016). Blood-brain barrier disruption induced by chronic sleep loss: Low-grade inflammation may be the link. *Journal of Immunology Research*, *2016* 4576012 [Epub 2016 Sep 21. Review].

Jacobsen, J. P. R., Krystal, A. D., Krishnan, K. R. R., & Caron, M. G. (2016). Adjunctive 5-Hydroxytryptophan slow-release for treatment-resistant depression: Clinical and preclinical rationale. *Trends in Pharmacological Sciences*, *37*(11), 933–944. https://doi.org/10.1016/j.tips.2016.09.001 [Epub 2016 Sep 28].

Jacobsen, J. P., Rudder, M. L., Roberts, W., Royer, E. L., Robinson, T. J., Oh, A., et al. (2016). SSRI augmentation by 5-hydroxytryptophan slow release: Mouse pharmacodynamic proof of concept. *Neuropsychopharmacology*, *41*(9), 2324–2334. https://doi.org/10.1038/npp.2016.35 [Epub 2016 Mar 2].

Johnson, I. S., Armstrong, J. G., Gorman, M., & Burnett, J. P., Jr. (1963). The vinca alkaloids: A new class of oncolytic agents. *Cancer Research*, *23*, 1390–1427.

Kaneko, M., Kumashiro, H., Takahashi, Y., & Hoshino, Y. (1979). L-5HTP treatment and serum 5-HT level after L-5-HTP loading on depressed patients. *Neuropsychobiology*, *5*(4), 232–240.

Kayacan, Y., Yazar, H., Cerit, G., & Ghojebeigloo, B. E. (2019). A new oxidative stress indicator: Effect of 5-hydroxytryptophan on thiol-disulfide homeostasis in exercise. *Nutrition*, *63-64*, 114–119. https://doi.org/10.1016/j.nut.2019.01.013 [Epub ahead of print].

Kerry, R. J., & Jenner, F. A. (1962). A double blind crossover comparison of diazepam (Valium, Ro5-2807) with chlordiazepoxide (Librium) in the treatment of neurotic anxiety. *Psychopharmacologia*, *3*, 302–306 [No abstract available].

Kiyatkin, E. A., Brown, P. L., & Sharma, H. S. (2007). Brain edema and breakdown of the blood-brain barrier during methamphetamine intoxication: Critical role of brain hyperthermia. *The European Journal of Neuroscience*, *26*(5), 1242–1253.

Kiyatkin, E. A., & Sharma, H. S. (2009). Permeability of the blood-brain barrier depends on brain temperature. *Neuroscience*, *161*(3), 926–939. https://doi.org/10.1016/j.neuroscience.2009.04.004 [Epub 2009 Apr 9].

Kobayashi, H., Magnoni, M. S., Govoni, S., Izumi, F., Wada, A., & Trabucchi, M. (1985). Neuronal control of brain microvessel function. *Experientia*, *41*(4), 427–434 [Review].

Kraus, C., Castrén, E., Kasper, S., & Lanzenberger, R. (2017). Serotonin and neuroplasticity—-Links between molecular, functional and structural pathophysiology in depression. *Neuroscience and Biobehavioral Reviews*, 77, 317–326. https://doi.org/10.1016/j.neubiorev.2017.03.007 [Epub 2017 Mar 22].

Krstić, M. K., & Katusić, Z. S. (1982). Divergent effects of cyproheptadine and R 50 656, a 5-HT2 antagonist, on the cardiovascular response to 5-hydroxytryptamine in rats. *European Journal of Pharmacology*, *85*(2), 225–227.

Lee, S. Y., Lee, S. J., Han, C., Patkar, A. A., Masand, P. S., & Pae, C. U. (2013). Oxidative/nitrosative stress and antidepressants: Targets for novel antidepressants. *Progress in Neuro-Psychopharmacology & Biological Psychiatry*, *46*, 224–235. https://doi.org/10.1016/j.pnpbp.2012.09.008 [Epub 2012 Sep 26].

Leysen, J. E., Awouters, F., Kennis, L., Laduron, P. M., Vandenberk, J., & Janssen, P. A. (1981). Receptor binding profile of R 41 468, a novel antagonist at 5-HT2 receptors. *Life Sciences*, *28*(9), 1015–1022 [No abstract available].

MacKenzie, E. T., & Scatton, B. (1987). Cerebral circulatory and metabolic effects of perivascular neurotransmitters. *CRC Critical Reviews in Clinical Neurobiology*, *2*(4), 357–419 [Review].

Maes, M., Calabrese, J., Jayathilake, K., & Meltzer, H. Y. (1997). Effects of subchronic treatment with valproate on L-5-HTP-induced cortisol responses in mania: Evidence for increased central serotonergic neurotransmission. *Psychiatry Research*, *71*(2), 67–76.

Marcus, M. L., Busija, D. W., Bischof, C. J., & Heistad, D. D. (1981). Methods for measurement of cerebral blood flow. *Federation Proceedings*, *40*(8), 2306–2310.

Marcus, M. L., Heistad, D. D., Ehrhardt, J. C., & Abboud, F. M. (1977). Regulation of total and regional spinal cord blood flow. *Circulation Research*, *41*(1), 128–134 [No abstract available].

Martin, T. G. (1996). Serotonin syndrome. *Annals of Emergency Medicine, 28*(5), 520–526 [Review].

Meltzer, H. Y., Wiita, B., Tricou, B. J., Simonovic, M., Fang, V., & Manov, G. (1982). Effect of serotonin precursors and serotonin agonists on plasma hormone levels. *Advances in Biochemical Psychopharmacology, 34*, 117–139 [Review. No abstract available].

Montagne, A., Nation, D. A., Pa, J., Sweeney, M. D., Toga, A. W., & Zlokovic, B. V. (2016). Brain imaging of neurovascular dysfunction in Alzheimer's disease. *Acta Neuropathologica, 131*(5), 687–707. https://doi.org/10.1007/s00401-016-1570-0 [Epub 2016 Apr 1. Review].

Montagne, A., Zhao, Z., & Zlokovic, B. V. (2017). Alzheimer's disease: A matter of blood-brain barrier dysfunction? *The Journal of Experimental Medicine, 214*(11), 3151–3169. https://doi.org/10.1084/jem.20171406 [Epub 2017 Oct 23. Review].

Musumeci, G., Castrogiovanni, P., Castorina, S., Imbesi, R., Szychlinska, M. A., Scuderi, S., et al. (2015). Changes in serotonin (5-HT) and brain-derived neurotrophic factor (BDFN) expression in frontal cortex and hippocampus of aged rat treated with high tryptophan diet. *Brain Research Bulletin, 119*(Pt. A), 12–18. https://doi.org/10.1016/j.brainresbull.2015.09.010Epub [2015 Oct 9].

Nakajima, T., Kudo, Y., & Kaneko, Z. (1978). Clinical evaluation of 5-hydroxy-L-tryptophan as an antidepressant drug. *Folia Psychiatrica et Neurologica Japonica, 32*(2), 223–230.

Nakamura, K., & Hasegawa, H. (2009). Production and peripheral roles of 5-HTP, a precursor of serotonin. *International Journal of Tryptophan Research, 2*, 37–43 [Epub 2009 Mar 30].

Nation, D. A., Sweeney, M. D., Montagne, A., Sagare, A. P., D'Orazio, L. M., Pachicano, M., et al. (2019). Blood-brain barrier breakdown is an early biomarker of human cognitive dysfunction. *Nature Medicine, 25*(2), 270–276. https://doi.org/10.1038/s41591-018-0297-y [Epub 2019 Jan 14].

Olsson, Y., Sharma, H. S., Pettersson, A., & Cervos-Navarro, J. (1992). Release of endogenous neurochemicals may increase vascular permeability, induce edema and influence cell changes in trauma to the spinal cord. *Progress in Brain Research, 91*, 197–203.

Ozkizilcik, A., Sharma, A., Muresanu, D. F., Lafuente, J. V., Tian, Z. R., Patnaik, R., et al. (2018). Timed release of cerebrolysin using drug-loaded titanate nanospheres reduces brain pathology and improves behavioral functions in Parkinson's disease. *Molecular Neurobiology, 55*(1), 359–369. https://doi.org/10.1007/s12035-017-0747-4.

Parsons, A. A. (1991). 5-HT receptors in human and animal cerebrovasculature. *Trends in Pharmacological Sciences, 12*(8), 310–315 [Review].

Pettersson, C. A., Sharma, H. S., & Olsson, Y. (1990). Vascular permeability of spinal nerve roots. A study in the rat with Evans blue and lanthanum as tracers. *Acta Neuropathologica, 81*(2), 148–154.

Rapoport, S. I. (1976). *Blood-brain barrier in physiology and medicine*. New York: Raven Press.

Raskin, N. H. (1981). Pharmacology of migraine. *Annual Review of Pharmacology and Toxicology, 21*, 463–478 [Review. No abstract available].

Reiche, R., & Frey, H. H. (1983). Antagonism of the 5-HT-induced bronchoconstriction in the cat. *Archives Internationales de Pharmacodynamie et de Thérapie, 263*(1), 139–145.

Reiche, S., Hermle, L., Gutwinski, S., Jungaberle, H., Gasser, P., & Majić, T. (2018). Serotonergic hallucinogens in the treatment of anxiety and depression in patients suffering from a life-threatening disease: A systematic review. *Progress in Neuro-Psychopharmacology & Biological Psychiatry, 81*, 1–10. https://doi.org/10.1016/j.pnpbp.2017.09.012 [Epub 2017 Sep 22].

Reyes-Gonzales, M. C., Fuentes-Broto, L., Martínez-Ballarín, E., Miana-Mena, F. J., Berzosa, C., García-Gil, F. A., et al. (2009). Effects of tryptophan and 5-hydroxytryptophan on the hepatic cell membrane rigidity due to oxidative stress. *The Journal of Membrane Biology, 231*(2–3), 93–99. https://doi.org/10.1007/s00232-009-9208-y [Epub 2009 Oct 22].

Sandyk, R. (2006). Serotonergic mechanisms in amyotrophic lateral sclerosis. *The International Journal of Neuroscience*, *116*(7), 775–826 [Review].

Saxena, P. R., Bom, A. H., & Verdouw, P. D. (1989). Characterization of 5-hydroxytryptamine receptors in the cranial vasculature. *Cephalalgia*, *9*(Suppl. 9), 15–22 [Review].

Sharif, Y., Jumah, F., Coplan, L., Krosser, A., Sharif, K., & Tubbs, R. S. (2018). Blood brain barrier: A review of its anatomy and physiology in health and disease. *Clinical Anatomy*, *31*(6), 812–823. https://doi.org/10.1002/ca.23083 [Epub 2018 Oct 18. Review].

Sharma, H. S. (1982). *Blood-brain barrier in stress*. [Ph D Thesis]. Varanasi, India: Banaras Hindu University Press.

Sharma, H. S. (1987). Effect of captopril (a converting enzyme inhibitor) on blood-brain barrier permeability and cerebral blood flow in normotensive rats. *Neuropharmacology*, *26*(1), 85–92.

Sharma, H. S. (1999). Pathophysiology of blood-brain barrier, brain edema and cell injury following hyperthermia: New role of heat shock protein, nitric oxide and carbon monoxide. an experimental study in the rat using light and electron microscopy. *Acta Universitatis Upsaliensis*, *830*, 1–94.

Sharma, H. S. (2000). Serotonin: From molecule to clinic, 2–3 November 2000, New Orleans, LA, USA. In *Current drugs, ID weekly highlights, 2000* (pp. 28–31). ISSN: 1365-5884, London UK. http://www.current-drugs.com.

Sharma, H. S. (2004). Influence of serotonin on the blood-brain and blood-spinal cord barriers. In H. S. Sharma & J. Westman (Eds.), *The blood-spinal cord and brain barriers in health and disease* (pp. 117–158). San Diego: Elsevier Academic Press.

Sharma, H. S. (2007). Methods to produce hyperthermia-induced brain dysfunction. *Progress in Brain Research*, *162*, 173–199 [Review].

Sharma, H. S. (2009). Blood–central nervous system barriers: The gateway to neurodegeneration, neuroprotection and neuroregeneration. In A. Lajtha, N. Banik, & S. K. Ray (Eds.), *Handbook of neurochemistry and molecular neurobiology: Brain and spinal cord trauma* (pp. 363–457). Berlin, Heidelberg, New York: Springer Verlag.

Sharma, H. S., Ali, S. F., Hussain, S. M., Schlager, J. J., & Sharma, A. (2009). Influence of engineered nanoparticles from metals on the blood-brain barrier permeability, cerebral blood flow, brain edema and neurotoxicity. An experimental study in the rat and mice using biochemical and morphological approaches. *Journal of Nanoscience and Nanotechnology*, *9*(8), 5055–5072.

Sharma, H. S., Ali, S. F., Tian, Z. R., Hussain, S. M., Schlager, J. J., Sjöquist, P. O., et al. (2009). Chronic treatment with nanoparticles exacerbate hyperthermia induced blood-brain barrier breakdown, cognitive dysfunction and brain pathology in the rat. Neuroprotective effects of nanowired-antioxidant compound H-290/51. *Journal of Nanoscience and Nanotechnology*, *9*(8), 5073–5090.

Sharma, H. S., Alm, P., & Westman, J. (1998). Nitric oxide and carbon monoxide in the brain pathology of heat stress. *Progress in Brain Research*, *115*, 297–333 [Review. No abstract available].

Sharma, H. S., Badgaiyan, R. D., Alm, P., Mohanty, S., & Wiklund, L. (2005). Neuroprotective effects of nitric oxide synthase inhibitors in spinal cord injury-induced pathophysiology and motor functions: An experimental study in the rat. *Annals of the New York Academy of Sciences*, *1053*, 422–434.

Sharma, H. S., Castellani, R. J., Smith, M. A., & Sharma, A. (2012). The blood-brain barrier in Alzheimer's disease: Novel therapeutic targets and nanodrug delivery. *International Review of Neurobiology*, *102*, 47–90. https://doi.org/10.1016/B978-0-12-386986-9.00003-X [Review].

Sharma, H. S., & Cervós-Navarro, J. (1990). Brain oedema and cellular changes induced by acute heat stress in young rats. *Acta Neurochirurgica. Supplementum (Wien)*, *51*, 383–386.

Sharma, H. S., Cervós-Navarro, J., & Dey, P. K. (1991a). Increased blood-brain barrier permeability following acute short-term swimming exercise in conscious normotensive young rats. *Neuroscience Research, 10*(3), 211–221.

Sharma, H. S., Cervós-Navarro, J., & Dey, P. K. (1991b). Acute heat exposure causes cellular alteration in cerebral cortex of young rats. *Neuroreport, 2*(3), 155–158.

Sharma, H. S., & Dey, P. K. (1981). Impairment of blood-brain barrier by immobilization stress: Role of serotonin. *Indian Journal of Physiology and Pharmacology, 25*, 111–122.

Sharma, H. S., & Dey, P. K. (1984). Role of 5-HT on increased permeability of blood-brain barrier under heat stress. *Indian Journal of Physiology and Pharmacology, 28*(4), 259–267.

Sharma, H. S., & Dey, P. K. (1986a). Probable involvement of 5-hydroxytryptamine in increased permeability of blood-brain barrier under heat stress in young rats. *Neuropharmacology, 25*(2), 161–167.

Sharma, H. S., & Dey, P. K. (1986b). Influence of long-term immobilization stress on regional blood-brain barrier permeability, cerebral blood flow and 5-HT level in conscious normotensive young rats. *Journal of the Neurological Sciences, 72*(1), 61–76.

Sharma, H. S., & Dey, P. K. (1987). Influence of long-term acute heat exposure on regional blood-brain barrier permeability, cerebral blood flow and 5-HT level in conscious normotensive young rats. *Brain Research, 424*(1), 153–162.

Sharma, H. S., & Dey, P. K. (1988). EEG changes following increased blood-brain barrier permeability under long-term immobilization stress in young rats. *Neuroscience Research, 5*(3), 224–239.

Sharma, H. S., Dey, P. K., & Olsson, Y. (1989). Brain edema, blood-brain barrier permeability and cerebral blood flow changes following intracarotid infusion of serotonin: Modification with cyproheptadine and indomethacin. In J. Krieglstein (Ed.), *Pharmacology of cerebral ischemia 1988* (pp. 317–323). Boca Raton, FL: CRC Press.

Sharma, H. S., Gordh, T., Wiklund, L., Mohanty, S., & Sjöquist, P. O. (2006). Spinal cord injury induced heat shock protein expression is reduced by an antioxidant compound H-290/51. An experimental study using light and electron microscopy in the rat. *Journal of Neural Transmission (Vienna), 113*(4), 521–536.

Sharma, H. S., & Kiyatkin, E. A. (2009). Rapid morphological brain abnormalities during acute methamphetamine intoxication in the rat: An experimental study using light and electron microscopy. *Journal of Chemical Neuroanatomy, 37*(1), 18–32. https://doi.org/10.1016/j.jchemneu.2008.08.002 [Epub 2008 Aug 19].

Sharma, H. S., Kiyatkin, E. A., Patnaik, R., Lafuente, J. V., Muresanu, D. F., Sjöquist, P. O., et al. (2015). Exacerbation of methamphetamine neurotoxicity in cold and hot environments: Neuroprotective effects of an antioxidant compound H-290/51. *Molecular Neurobiology, 52*(2), 1023–1033. https://doi.org/10.1007/s12035-015-9252-9 [Epub 2015 Jun 26].

Sharma, H. S., Miclescu, A., & Wiklund, L. (2011). Cardiac arrest-induced regional blood-brain barrier breakdown, edema formation and brain pathology: A light and electron microscopic study on a new model for neurodegeneration and neuroprotection in porcine brain. *Journal of Neural Transmission (Vienna), 118*(1), 87–114. https://doi.org/10.1007/s00702-010-0486-4 [Epub 2010 Oct 21].

Sharma, H. S., Muresanu, D. F., Lafuente, J. V., Nozari, A., Patnaik, R., Skaper, S. D., et al. (2016). Pathophysiology of blood-brain barrier in brain injury in cold and hot environments: Novel drug targets for neuroprotection. *CNS & Neurological Disorders Drug Targets, 15*(9), 1045–1071 [Review].

Sharma, A., Muresanu, D. F., Ozkizilcik, A., Tian, Z. R., Lafuente, J. V., Manzhulo, I., et al. (2019). Sleep deprivation exacerbates concussive head injury induced brain pathology: Neuroprotective effects of nanowired delivery of cerebrolysin with α-melanocyte-stimulating hormone. *Progress in Brain Research, 245*, 1–55. https://doi.org/10.1016/bs.pbr.2019.03.002 [Epub 2019 Apr 2].

Sharma, H. S., Nyberg, F., Cervos-Navarro, J., & Dey, P. K. (1992). Histamine modulates heat stress-induced changes in blood-brain barrier permeability, cerebral blood flow, brain oedema and serotonin levels: An experimental study in conscious young rats. *Neuroscience*, *50*(2), 445–454.

Sharma, H. S., Nyberg, F., Westman, J., Alm, P., Gordh, T., & Lindholm, D. (1998). Brain derived neurotrophic factor and insulin like growth factor-1 attenuate upregulation of nitric oxide synthase and cell injury following trauma to the spinal cord. An immunohistochemical study in the rat. *Amino Acids*, *14*(1–3), 121–129.

Sharma, H. S., & Olsson, Y. (1990). Edema formation and cellular alterations following spinal cord injury in the rat and their modification with p-chlorophenylalanine. *Acta Neuropathologica*, *79*(6), 604–610.

Sharma, H. S., Olsson, Y., & Cervós-Navarro, J. (1993). Early perifocal cell changes and edema in traumatic injury of the spinal cord are reduced by indomethacin, an inhibitor of prostaglandin synthesis. Experimental study in the rat. *Acta Neuropathologica*, *85*(2), 145–153.

Sharma, H. S., Olsson, Y., & Dey, P. K. (1989). Influence of serotonin on regional blood-brain barrier permeability and cerebral blood flow in the rat. In I. J. Seylaz & E. T. MacKenzie (Eds.), *Neurotransmission and cerebrovascular function* (pp. 253–255). Amsterdam: Elsevier.

Sharma, H. S., Olsson, Y., & Dey, P. K. (1990a). Blood-brain barrier permeability and cerebral blood flow following elevation of circulating serotonin level in the anaesthetized rats. *Brain Research*, *517*, 215–223.

Sharma, H. S., Olsson, Y., & Dey, P. K. (1990b). Early accumulation of serotonin in rat spinal cord subjected to traumatic injury. Relation to edema and blood flow changes. *Neuroscience*, *36*(3), 725–730.

Sharma, H. S., Olsson, Y., & Dey, P. K. (1995). Serotonin as a mediator of increased microvascular permeability of the brain and spinal cord. Experimental observations in anaesthetised rats and mice. In J. Greenwood, D. Begley, M. Segal, & S. Lightman (Eds.), *New concepts of a blood-brain barrier* (pp. 75–80). New York: Plenum Press.

Sharma, H. S., Olsson, Y., Persson, S., & Nyberg, F. (1995). Trauma-induced opening of the blood-spinal cord barrier is reduced by indomethacin, an inhibitor of prostaglandin biosynthesis. Experimental observations in the rat using [131I]-sodium, Evans blue and lanthanum as tracers. *Restorative Neurology and Neuroscience*, 7(4), 207–215. https://doi.org/10.3233/RNN-1994-7403.

Sharma, H. S., Patnaik, R., Sharma, A., Lafuente, J. V., Miclescu, A., & Wiklund, L. (2015). Cardiac arrest alters regional ubiquitin levels in association with the blood-brain barrier breakdown and neuronal damages in the porcine brain. *Molecular Neurobiology*, *52*(2), 1043–1053. https://doi.org/10.1007/s12035-015-9254-7 [Epub 2015 Jun 25].

Sharma, H. S., & Sjöquist, P. O. (2002). A new antioxidant compound H-290/51 modulates glutamate and GABA immunoreactivity in the rat spinal cord following trauma. *Amino Acids*, *23*(1–3), 261–272.

Sharma, H. S., & Westman, J. (2003). Blood-spinal cord and brain barriers in health and disease. In *Preface*: *Vol. xiii–xv*. San Diego, Boston: Elsevier Academic Press.

Sharma, H. S., & Westman, J. (2004). *The blood-spinal cord and brain barriers in health and disease* (pp. 1–617). San Diego: Academic Press [Release date: Nov. 9, 2003].

Sharma, H. S., Westman, J., Cervós-Navarro, J., Dey, P. K., & Nyberg, F. (1995). Probable involvement of serotonin in the increased permeability of the blood-brain barrier by forced swimming. An experimental study using Evans blue and ^{131}I-sodium tracers in the rat. *Behavioural Brain Research*, *72*, 189–196.

Sharma, H. S., Westman, J., Cervós-Navarro, J., Dey, P. K., & Nyberg, F. (1997). Opioid receptor antagonists attenuate heat stress-induced reduction in cerebral blood flow, increased blood-brain barrier permeability, vasogenic edema and cell changes in the rat. *Annals of the New York Academy of Sciences*, *813*, 559–571 [No abstract available].

Sharma, H. S., Westman, J., Cervós-Navarro, J., & Nyberg, F. (1997). Role of neurochemicals in brain edema and cell changes following hyperthermic brain injury in the rat. *Acta Neurochirurgica. Supplement, 70*, 269–274.

Sharma, H. S., Westman, J., & Nyberg, F. (1998). Pathophysiology of brain edema and cell changes following hyperthermic brain injury. *Progress in Brain Research, 115*, 351–412 [Review. No abstract available].

Sharma, H. S., Westman, J., Nyberg, F., Cervos-Navarro, J., & Dey, P. K. (1994). Role of serotonin and prostaglandins in brain edema induced by heat stress. An experimental study in the young rat. *Acta Neurochirurgica. Supplementum (Wien), 60*, 65–70.

Sharma, H. S., Westman, J., Olsson, Y., & Alm, P. (1996). Involvement of nitric oxide in acute spinal cord injury: An immunocytochemical study using light and electron microscopy in the rat. *Neuroscience Research, 24*(4), 373–384.

Sharpe, G. L., Thalme, B., & Larsson, K. S. (1974). Studies on closure of the ductus arteriosus. XI. Ductal closure in utero by a prostaglandin synthetase inhibitor. *Prostaglandins, 8*(5), 363–368 [No abstract available].

Shelton, R. C. (2019). Serotonin and norepinephrine reuptake inhibitors. *Handbook of Experimental Pharmacology, 250*, 145–180. https://doi.org/10.1007/164_2018_164 [Epub ahead of print].

Sicuteri, F., Anselmi, B., & Fanciullacci, M. (1970). Parachlorophenylalanine, an inhibitor of enteramine (serotonin) synthesis in migraine. *Research and Clinical Studies in Headache, 3*, 310–314 [No abstract available].

Sinopoli, V. M., Burton, C. L., Kronenberg, S., & Arnold, P. D. (2017). A review of the role of serotonin system genes in obsessive-compulsive disorder. *Neuroscience and Biobehavioral Reviews, 80*, 372–381. https://doi.org/10.1016/j.neubiorev.2017.05.029 [Epub 2017 May 30].

Snyder, S. H., Axelrod, J., & Zweig, M. (1965). A sensitive and specific fluorescence assay for tissue serotonin. *Biochemical Pharmacology, 14*(5), 831–835.

Sweeney, M. D., Kisler, K., Montagne, A., Toga, A. W., & Zlokovic, B. V. (2018). The role of brain vasculature in neurodegenerative disorders. *Nature Neuroscience, 21*(10), 1318–1331. https://doi.org/10.1038/s41593-018-0234-x [Epub 2018 Sep 24. Review].

Sweeney, M. D., Montagne, A., Sagare, A. P., Nation, D. A., Schneider, L. S., Chui, H. C., et al. (2019). Vascular dysfunction-The disregarded partner of Alzheimer's disease. *Alzheimer's & Dementia, 15*(1), 158–167. https://doi.org/10.1016/j.jalz.2018.07.222.

Sweeney, M. D., Sagare, A. P., & Zlokovic, B. V. (2018). Blood-brain barrier breakdown in Alzheimer disease and other neurodegenerative disorders. *Nature Reviews. Neurology, 14*(3), 133–150. https://doi.org/10.1038/nrneurol.2017.188 [Epub 2018 Jan 29. Review].

Sweeney, M. D., Zhao, Z., Montagne, A., Nelson, A. R., & Zlokovic, B. V. (2019). Blood-brain barrier: From physiology to disease and back. *Physiological Reviews, 99*(1), 21–78. https://doi.org/10.1152/physrev.00050.2017 [Review].

Takahashi, S., Takahashi, R., Masumura, I., & Miike, A. (1976). Measurement of 5-hydroxyindole compounds during L-5-HTP treatment in depressed patients. *Folia Psychiatrica et Neurologica Japonica, 30*(4), 463–473.

Udenfriend, S., Weissbach, H., & Bogdanski, D. F. (1957). Increase in tissue serotonin following administration of its precursor 5-hydroxytryptophan. *The Journal of Biological Chemistry, 224*(2), 803–810.

Vadodaria, K. C., Stern, S., Marchetto, M. C., & Gage, F. H. (2018). Serotonin in psychiatry: In vitro disease modeling using patient-derived neurons. *Cell and Tissue Research, 371*(1), 161–170. https://doi.org/10.1007/s00441-017-2670-4 [Epub 2017 Aug 15].

Van Nueten, J. M., Xhonneux, R., Vanhoutte, P. M., & Janssen, P. A. (1981). Vascular activity of ketanserin (R 41 468), a selective 5-HT2 receptor antagonist. *Archives Internationales de Pharmacodynamie et de Thérapie, 250*(2), 328–329 [No abstract available].

van Vliet, I. M., Slaap, B. R., Westenberg, H. G., & Den Boer, J. A. (1996). Behavioral, neuroendocrine and biochemical effects of different doses of 5-HTP in panic disorder. *European Neuropsychopharmacology*, *6*(2), 103–110.

Wiernsperger, N. (1990). Serotonin 5-HT2 receptors and brain circulation. *Journal of Cardiovascular Pharmacology*, *16*(Suppl. 3), S20–S24 [Review].

Winkler, T., Sharma, H. S., Stålberg, E., Olsson, Y., & Dey, P. K. (1995). Impairment of blood-brain barrier function by serotonin induces desynchronisation of spontaneous cerebral cortical activity. Experimental observations in the anaesthetized rat. *Neuroscience*, *68*, 1097–1104.

Yeung, M., Treit, D., & Dickson, C. T. (2012). A critical test of the hippocampal theta model of anxiolytic drug action. *Neuropharmacology*, *62*(1), 155–160. https://doi.org/10.1016/j.neuropharm.2011.06.011 [Epub 2011 Jun 23].

Yohn, C. N., Gergues, M. M., & Samuels, B. A. (2017). The role of 5-HT receptors in depression. *Molecular Brain*, *10*(1), 28. https://doi.org/10.1186/s13041-017-0306-y.

Zhou, Q. G., Zhu, X. H., Nemes, A. D., & Zhu, D. Y. (2018). Neuronal nitric oxide synthase and affective disorders. *IBRO Reports*, *5*, 116–132. https://doi.org/10.1016/j.ibror.2018.11.004 [eCollection 2018 Dec. Review].

Zmilacher, K., Battegay, R., & Gastpar, M. (1988). L-5-hydroxytryptophan alone and in combination with a peripheral decarboxylase inhibitor in the treatment of depression. *Neuropsychobiology*, *20*(1), 28–35.

CHAPTER TWO

Anesthetics influence concussive head injury induced blood-brain barrier breakdown, brain edema formation, cerebral blood flow, serotonin levels, brain pathology and functional outcome

Hari Shanker Sharma[a,*], **Dafin Fior Muresanu**[b,c], **Ala Nozari**[d], **Prasanta Kumar Dey**[e], **Lars Wiklund**[a], **Aruna Sharma**[a]

[a]International Experimental Central Nervous System Injury & Repair (IECNSIR), Department of Surgical Sciences, Anesthesiology & Intensive Care Medicine, Uppsala University Hospital, Uppsala University, Uppsala, Sweden
[b]Department of Clinical Neurosciences, University of Medicine & Pharmacy, Cluj-Napoca, Romania
[c]"RoNeuro" Institute for Neurological Research and Diagnostic, Cluj-Napoca, Romania
[d]Anesthesia and Critical Care, Massachusetts General Hospital, Boston, MA, United States
[e]Neurophysiology Research Unit, Department of Physiology, Institute of Medical Sciences, Banaras Hindu University, Varanasi, Uttar Pradesh, India
*Corresponding author: e-mail addresses: sharma@surgsci.uu.se; hssharma@aol.com; harishanker_sharma55@icloud.com

Contents

International Review of Neurobiology, Volume 146
ISSN 0074-7742
https://doi.org/10.1016/bs.irn.2019.06.006

Abstract

Several lines of evidences show that anesthetics influence neurotoxicity and neuroprotection. The possibility that different anesthetic agents potentially influence the pathophysiological and functional outcome following neurotrauma was examined in a rat model of concussive head injury (CHI). The CHI was produced by an impact of 0.224N on the right parietal bone by dropping a weight of 114.6g from a 20cm height under different anesthetic agents, e.g., inhaled ether anesthesia or intraperitoneally administered ketamine, pentobarbital, equithesin or urethane anesthesia. Five hour CHI resulted in profound volume swelling and brain edema formation in both hemispheres showing disruption of the blood-brain barrier (BBB) to Evans blue and radioiodine. A marked decrease in the cortical CBF and a profound increase in plasma or brain serotonin levels were seen at this time. Neuronal damages were present in several parts of the brain. These pathological changes were most marked in CHI under ether anesthesia followed by ketamine (35mg/kg, i.p.), pentobarbital (50mg/kg, i.p.), equithesin (3mL/kg, i.p.) and urethane (1g/kg, i.p.). The functional outcome on Rota Rod performances or grid walking tests was also most adversely affected after CHI under ether anesthesia followed by pentobarbital, equithesin and ketamine. Interestingly, the plasma and brain serotonin levels strongly correlated with the development of brain edema in head injured animals in relation to different anesthetic agents used. These observations suggest that anesthetic agents are detrimental to functional and pathological outcomes in CHI probably through influencing the circulating plasma and brain serotonin levels, not reported earlier. Whether anesthetics could also affect the efficacy of different neuroprotective agents in CNS injuries is a new subject that is currently being examined in our laboratory.

1. Introduction

General anesthesia that is used for thousands of patients worldwide daily was discovered almost 170 years ago (Bouroche & Bourgain, 2015; Hingson, 1961; Learned, 1949; Oliveira, Bernardo, & Nunes, 2017; Smith, Cozowicz, Uda, Memtsoudis, & Barrington, 2017; Troup, 1947; Watrous, Davis, & Anderson, 1950a, 1950b). However, the detailed physiological mechanisms of action with regard to anesthetics neurotoxicity or neuroprotection are still not well known (De Hert, Preckel, & Schlack, 2009; Dundee, 1967; Kopp Lugli, Yost, & Kindler, 2009; O'Brien, 1967; Roberts & Jagdish, 2016; Robinson & Toledo, 2012). Recent developments in electrophysiology and molecular biotechnology suggest the prime targets of anesthetics are not only lipid membranes but also receptors and ion channels (Halsey, 1992; Narahashi, 1991; Pocock & Richards, 1993). Anesthetics although have reversible actions but control immobility, amnesia and unconsciousness with analgesia, muscle relaxation, suppression of reflexes are the prerequisites for successful surgeries in patients (Eger & Sonner, 2006; Villars, Kanusky, & Dougherty, 2004). However, most of the anesthetics lead to long-term effect on postoperative cognitive dysfunction (seeDe Hert et al., 2009; Kopp Lugli et al., 2009; O'Brien, 1967). However, none of the anesthetics used today have all these characteristics in one drug used for general anesthesia (De Hert et al., 2009; Kopp Lugli et al., 2009).

There are evidences that various kinds of anesthetics affect consciousness at different levels in the brain (Alkire, Hudetz, & Tononi, 2008; Cascella, Bimonte, & Muzio, 2018; Ferreira, Nunes, Mendes, & Amorim, 2019; Sepúlveda, Tapia, & Monsalves, 2019). In general the inhaled anesthetics such as halothane and isoflurane produce a global decrease in glucose metabolism in the brain at the level where consciousness is lost (Park et al., 2017; Sano et al., 2016). On the other hand intravenous anesthetic propofol lowers glucose metabolism in subcortical regions such as medial thalamus and related arousal brain areas as compared to the cerebral cortex (Laaksonen et al., 2018; Manell, Jensen-Waern, & Hedenqvist, 2017). These observations support the idea that anesthetics interfere with corticothalamic oscillations to achieve unconsciousness (De Hert et al., 2009; Kopp Lugli et al., 2009; Soplata et al., 2017).

At the cellular level anesthetics induce their effects either by promoting inhibition or diminishing excitation in various parts of the brain (Brohan & Goudra, 2017;Kopp Lugli et al., 2009; Pocock & Richards, 1993). Thus, the main effect of anesthetics lies on synapses either at presynaptic region or the post synaptic area (Dundee, 1967; Kopp Lugli et al., 2009; Musizza & Ribaric, 2010; O'Brien, 1967; Wali, 1985)- A presynaptic effect of anesthetics will alter release of the neurotransmitters, whereas a postsynaptic action lead to changes in the amplitude or frequency of impulses that traverse the synapse (Dundee, 1967; Kopp Lugli et al., 2009; O'Brien, 1967). In general, the main effect of anesthetics is largely due to a change in the balance between inhibition and excitation of neural transmission within the CNS (see Kopp Lugli et al., 2009). Studies in spinal cord showed that anesthetics impair excitatory transmission more intensely than the potentiation of inhibitory transmission (Kopp Lugli et al., 2009; Pocock & Richards, 1993).

Regarding molecular level of anesthetics action, the chloride channels gamma-aminobutyric acid (GABA-a and glycine receptors together with potassium channels K2P, Kv and K-ATP are the main inhibitory ion channels for their action (De Hert et al., 2009; Kopp Lugli et al., 2009; Pocock & Richards, 1993). Anesthetics induced inhibition of excitatory ion channels include acetylcholine nicotinic and muscarinic receptors, kainate and *N*-methyl-D-aspartate (NMDA) receptors together with serotonin 5-HT2 and 5-HT3 receptors for their specific anesthetic effects (Kopp Lugli et al., 2009; Pocock & Richards, 1993).

1.1 Anesthetics effects on neuroprotection and neurotoxicity

Although, anesthetic effects on unconsciousness appears to be quickly and completely reversible, in recent years great concern has been raised over long-term effects on neurotoxicity, neurodegeneration and functional cognitive dysfunction (Alam et al., 2017; Bilotta, Evered, & Gruenbaum, 2017; Clausen, Kähler, & Hansen, 2018; Disma et al., 2018; Van Biesen, Van de Velde, & Rex, 2015; Vutskits & Davidson, 2017; Wu, Zhao, Weng, & Ma, 2019). Also anesthetic effects of neuro and cardio protection are equally highlighted (Gardner & Menon, 2018; Klamt et al., 2017; Kunst & Klein, 2015; Marchesini & Disma, 2019; Pasternak & Lanier, 2007; Slupe & Kirsch, 2018; Wink, Veering, Aarts, & Wouters, 2019; Xia, Li, & Irwin, 2016).

Neuroproetction by anesthetics is largely due to their ability to inhibit GABA-a inhibitory and NMDA excitatory receptors (Muir, 2010; Olsen & Li, 2011; Petrenko, Yamakura, Sakimura, & Baba, 2014; Saari, Uusi-Oukari, Ahonen, & Olkkola, 2011; Son, 2010; Trapani, Altomare, Liso, Sanna, & Biggio, 2000; Yan & Jiang, 2014). Activation of K2P potassium channel could also protect neurons from hypoxic and glucose deprivation injury (Franks & Honoré, 2004; Riegelhaupt, Tibbs, & Goldstein, 2018; Steinberg, Wafford, Brickley, Franks, & Wisden, 2015).

Since anesthetics disrupt mental processes exposure to anesthetic agents can induce neurodegenerative changes in young or infant brain (Andropoulos, 2018; Dubey, Nau, & Sycip, 2019). Dramatic neurodegeneration in hippocampal neurons were reported following anesthetic exposure during the first postnatal week in the rodent brain (Andropoulos, 2018; Dubey et al., 2019; Goyagi, 2019; Jevtovic-Todorovic et al., 2003; Li et al., 2018). In this regards several anesthetics such as isoflurane, ketamine and propofol induces neuronal injury in young or developing brain (see Petrenko et al., 2014; Trapani et al., 2000; Yan & Jiang, 2014). The main reasons of neurodegeneration by anesthetics in young brain are largely due to blockade of excitatory and potentiation of inhibitory ion channels (Eger & Sonner, 2006; Kopp Lugli et al., 2009; Pocock & Richards, 1993; Villars et al., 2004). In children who are exposed to two or more episodes of anesthetic exposure in their first 2 years of life is associated with their learning disabilities in later life (Dubey et al., 2019; Jevtovic-Todorovic et al., 2003; Pasternak & Lanier, 2007).

Whether different anesthetics can also affect pathophysiology of neurotrauma and therapeutic aspects is still unclear. In this investigation, we have explored influence of various widely used anesthetics in experimental models to study outcome of concussive head injury (CHI) with regard to pathophysiology and functional outcome.

2. Methodological considerations

2.1 Animals

Experiments were carried out on male Wistar rats (250–300 g body weight) housed at controlled ambient temperature ($21 \pm 1\,^{\circ}C$) with 12 h light and 12 h dark schedule. The food and tap water were supplied ad libitum before the experiment. All experiments were conducted according to National Institute of Health Guidelines for Handling and Care of Experimental animals and approved by local Institutional Ethics Committee.

2.2 Concussive head injury

Under different anesthetic agents (see below) the skull bone was opened and made free from scalp muscles. The head of the animal was fixed in stereotaxic apparatus so that it would not move following any impact (Dey & Sharma, 1984; Sharma et al., 2007a, 2010; Vannemreddy et al., 2006a). Using an aluminum guide tube, a weight of 114.6 g was dropped over the right parietal bone from a 20 cm height (see Fig. 1). This experimental condition results in an impact of 0.224 N over the skull and induces profound concussive injury. In few animals (<0.5%) the skull bone was slightly fractured and those animals were discarded. Uninjured animals under identical conditions and same anesthesia were used as controls (Dey & Sharma, 1984; Sharma et al., 2007a, 2010).

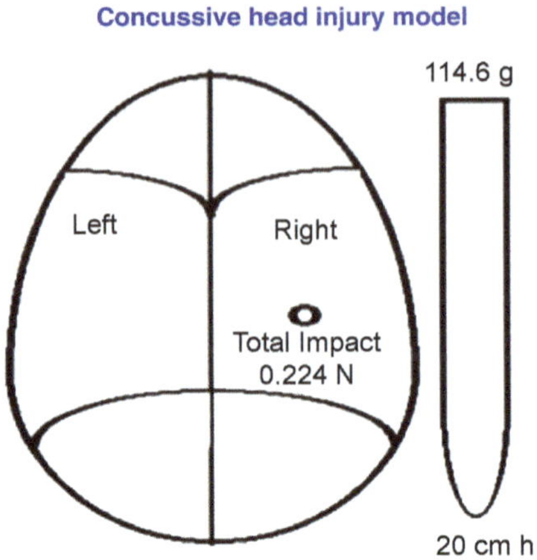

Fig. 1 Concussive head injury (CHI) model in the rat. Under anesthesia, a metal rod weighing 114.6g slightly tapered at the end was dropped over the right parietal skull (◦) from 20cm height through an aluminum guide tube. The rat head is properly cushioned and fixed in a stereotaxic frame to prevent to prevent any kind of rebound movement. This weight and height yield an impact force of 0.224N on the injury point over the skull. For details see text. *Data modified after Sharma, H. S. (2004) Blood-brain and spinal cord barriers in stress. In H. S. Sharma and J., & Westman (Eds)* The Blood-Spinal Cord and Brain Barriers in Health and Disease *(pp. 231–298). San Diego: Elsevier Academic Press.*

2.3 Anesthetic agents

Rats were anesthetized with different anesthetic agents and CHI was induced. All these animals were allowed to survive 5 h after injury. The anesthetic agents used were as follows:

2.3.1. Diethyl ether (British Drug House, England) as inhalation anesthesia. For this purpose, rats were allowed to breath a mixture of ether anesthesia and fresh air until the rats reached anesthesia level IV (Bourne, 1946; Bovill, 2008; Defalque & Wright, 2008; England & Jones, 1992; Finer, 1965; Greuter, Weil, Hildebrandt, & Surbeck, 2017; No Author Listed, 1852).

2.3.2. Pentobarbital (Nembutal, Abbott Laboratories, UK) was administered in rats at a dose of 50mg/kg, i.p. Prior to pentobarbital anesthesia, animals were treated with Atropine sulfate (Sigma, USA, 1 mg/kg, s.c.) to minimize bronchial secretion, if any in animals during the surgery and revival (Belopavlovic & Buchthal, 1980; Daniel, Fry, & Savord, 1984; Lowenstein, Aminoff, & Simon, 1988; Malviya et al., 2004; Okamoto, 1984; Zgleszewski, Zurakowski, Fontaine, D'Angelo, & Mason, 2008).

2.3.3. Equithesin (Umeå, Sweden) is a mixture of various anesthetic agents in suitable proportions (Deacon & Rawlins, 1996; Gaese & Ostwald, 2001). Equithesin was administered in rats intraperitoneally at the dose of 3 mL/kg to attain grade IV of surgical anesthesia.

2.3.4. Urethane (Merck, Germany) was administered intraperitoneally in the rats at a dose of 1.5 g/kg (Field & Lang, 1988; Maggi & Meli, 1986a, 1986b, 1986c; Pagliardini, Funk, & Dickson, 2013). This is an irreversible anesthesia and thus functional outcome in these animals could not be monitored.

2.4 Parameters measured

The following parameters were measured in control and head injured animals.

2.4.1 Physiological variables

Mean arterial blood pressure (MABP), arterial pH and blood gases were analyzed to record these physiological variables in CHI (Sharma, 1987; Sharma & Dey, 1986a). For MABP recordings under anesthesia a polyethylene catheter PE 10 was inserted into the right common carotid artery retrogradely towards heart and secured there. The other end of the catheter was connected to a Statham P23 pressure transducer (USA) to record MABP (Sharma & Dey, 1986b). The output was fed to a chart recorder (Electromed, UK). Immediately before connecting the arterial catheter to pressure transducer, about 1 mL arterial blood sample was withdrawn for later determination of arterial pH and $PaCO_2$ and PaO_2 using a Radiometer apparatus (Copenhagen, Denmark).

2.4.2 Functional outcome

The functional outcome was measured in animals that received Equithesin, diethyl ether, ketamine and pentobarbital using Rota Rod performances and grid walking tests.

(a) Rota Rod

Animals were placed on a Rota Rod maintained at 16 rpm on which normal animals could stand without any problems for 120 s. However, animals with head trauma could not stay that longer. The duration of stay on the Rota Rod was counted manually for each animal in various groups to assess the cognitive function of the rats after head injury (see Sharma, 2006; Sharma & Sharma, 2007).

(b) Grid walking test

Animals were also subjected to grid-walking test on a stainless steel wire mesh (size 30 mm) elevated at 30° on which rats were placed to walk for about 1 m distance (Sharma, 2006). Normal rats could walk freely and the number of steps taken was counted during 60 s period. The injured animals walk slowly and also not able to place their fore-limbs correctly on the mesh and the limbs are often misplaced into the grid. The number of placement errors were manually counted for each animal and calculated as percentage errors (Sharma, 2006; Sharma et al., 2009; Sharma & Sharma, 2007).

2.4.3 Brain edema

After the end of the experiments, animals were decapitated and the cerebral cortical areas of the inured and non-injured side were taken out. Large superficial vessels and any blood clot were removed quickly (Sharma, 2007). The samples were weighted immediately and placed in an incubator maintained at 90 °C for 72 h to obtain dry weight (Sharma, Westman, & Nyberg, 1998). The water content was calculated form the difference between wet and dry weight of the samples (Sharma, 2007;Sharma et al., 1998; Sharma & Cervós-Navarro, 1990).

2.4.4 Blood-brain barrier

Leakage of blood-brain barrier to proteins in the cortical areas was examined using Evans blue albumin and $^{[131]}$Iodine extravasations (Sharma, 1987; Sharma & Dey, 1986a, 1986b). Both terraces when introduced into the circulation will bind to serum albumin (Sharma & Dey, 1986b). Thus, a leakage of these tracers in the brain represent extravasation of serum protein compels (Sharma, 1987, 2007; Sharma et al., 1998; Sharma & Dey, 1986a, 1986b). These tracers were administered intravenously 5 min before termination of the experiments. After the end of the experiments, the intravascular tracer was washed out through transcardiac profusion. After saline perfusion, the brain was dissected out and the cortical areas form both hemispheres were removed and weighed immediately (Sharma et al., 1998; Sharma & Cervós-Navarro, 1990). Leakage of blue dye was examined and then the samples were counted in a 3-in well type Gamma counter (Packard, United States). Immediately before perfusion a sample of whole blood was withdrawn from the heart for the measurement of blood radioactivity. The leakage of radiotracer in the brain was expressed as percentage increase over blood radioactivity (Sharma, 1987; Sharma & Dey, 1986a). After counting the radioactivity, the cortical tissues were homogenized in a mixture of sodium sulfate and acetone to extract the Evans blue dye in solution (Sharma & Dey, 1986b). The Evans blue dye entered into brain samples were measured in a spectrophotometer (Carl Zeiss, Germany) at 620 nm. The concentration of dye was calculated from the standard Evans blue solution (Sharma, 1987).

2.4.5 Cerebral blood flow

Cortical cerebral blood flow (CBF) was measured using [125]Iodine labeled carbonized microspheres (15 ⊥ 0.6 μm o.d.) as described earlier (Sharma & Dey, 1988; Sharma, Olsson, & Dey, 1990). In brief, about 1 million microspheres were administered into the common carotid artery through a cannula retrogradely placed near the heart within 90 s (Sharma, 1987). Timed arterial blood samples from the right femoral artery was withdrawn at every 30 s starting from 30 s before start of the radiotracer administration and continued up to 30 s after completion of the microsphere injection at the rate of 0.8 mL/min (Sharma, 1987; Sharma & Dey, 1986a, 1986b). At the end of the experiment, the rats were decapitated and the brains ad taken out. The large superficial blood vessels and blood clots, if any were removed and the cortical samples form the right and left hemispheres were dissected out, weighed and the radioactivity was counted in a 3-in well type Gamma counter (Packard, USA) (Sharma, 1987). Total count of blood radioactivity was also made in serial blood samples collected over time (Sharma, 1987; Sharma & Dey, 1986b). The blood flow was calculated from the amount of radioactivity present in the cortical brain region at the time of killing in reference to the arterial blood radioactivity multiplied by the rate of withdrawal of the blood samples (see Sharma, 1987; Sharma et al., 1990; Sharma & Dey, 1986a, 1986b, 1988).

2.4.6 Serotonin levels

The plasma and brain serotonin were measured biochemically using Spectrophotoflurometer as described in details earlier (Dey & Sharma, 1983; Sharma & Dey, 1981). In brief, the arterial blood withdrawn from the rats along with cortical brain samples were deproteinized with 0.4% per chloric acid and the samples were processed from serotonin measurement using butanol extraction in a salt saturated alkaline medium (pH 10.0)

(Sharma & Dey, 1986a, 1986b). After purification of extract the serotonin fluorescence was developed using ninhydrin by heating the samples at 75 °C for 30 min. The florescence was measured in cooled samples at room temperature using an Aminco Bowman Spectrophotofluorometer (USA) at an excitation wavelength of 480 and emission wavelength 520 nm (Dey & Sharma, 1983, 1984; Sharma & Dey, 1981, 1986b).

2.4.7 Histopathological analysis of neuronal damage

For this purpose, separate groups of rats after 5 h head injury were perfused transcardially with 4% paraformaldehyde precede with a brief saline rinse (Sharma et al., 1998; Sharma & Cervós-Navarro, 1990). The brain tissues were dissected out and processed for paraffin embedding (Sharma, Cervós-Navarro, & Dey, 1991a; Sharma, Zimmer, Westman, & Cervós-Navarro, 1992). About 3-μm thick sections were cut and stained with Nissl or Hematoxylin or Eosin using standard protocol (Sharma et al., 1991a). The neuronal damage was examined under a bright field Leica Microscope and the image was recorded on a professional Kodak color negative film (100 ASA) and processed for digital images in Kodak Professional Laboratory, Stockholm (Sharma, Olsson, & Cervós-Navarro, 1993; Sharma & Sjöquist, 2002). The final images were processed in a Macintosh computer (System Mac Os 10.3.4) using Adobe Photoshop 7.0 employing identical color balance an filters for control and injured animals from all groups (Sharma & Sjöquist, 2002).

2.4.8 Statistical analysis

Quantitative data obtained were analyzed using ANOVA followed by Dunnet's test for multiple group comparison. Semiquantitative data (morphology and behavioral outcome) were analyzed using Chi-Square test. A P-value <0.05 were considered significant.

3. Our observations on anesthetics influence on concussive head injury

We observed marked differences in the physiological variables, BBB breakdown, CBF, edema formation and serotonin levels in plasma and brain following CHI under different anesthetic used. Differences in the functional outcome after CHI were also observed with different anesthetics used. The main findings are described below.

3.1 Physiological variables

Physiological variables in normal control animals were not significantly influenced by any anesthetic used in this investigation. Thus, changes in the MABP, arterial pH or blood gases PaO_2 or $PaCO_2$ were not significantly different in any group whether anesthetized with ether, ketamine, pentobarbital, Equithesin or urethane (see Table 1).

However, when CHI was inflicted under these different anesthetics significant changes in the MABP, arterial pH and blood gases were seen

Table 1 Effect of anesthetics on physiological variables in controls and closed head injury at 5 h.

Type of experiment	Mean arterial blood pressure (MABP) torr	Arterial pH	Arterial PaO_2 torr	Arterial $PaCO_2$ torr
A. Control				
Ether	108 ± 7	7.37 ± 0.04	80.34 ± 0.12	33.45 ± 0.21
Ketamine 35 mg/kg, i.p.	112 ± 6	7.38 ± 0.03	80.56 ± 0.19	33.54 ± 0.21
Pentobarbital 50 mg/kg, i.p.	110 ± 6	7.37 ± 0.06	80.45 ± 0.14	33.56 ± 0.21
Equithesin 3 mL/kg, i.p.	114 ± 6	7.38 ± 0.02	80.38 ± 0.18	33.44 ± 0.12
Urethane# 1 g/kg, i.p.	110 ± 6	7.38 ± 0.07	80.37 ± 0.14	33.42 ± 0.21
B. Concussive head injury (CHI)				
Ether inhalation	78 ± 5**	7.36 ± 0.03*	82.34 ± 0.23**	34.67 ± 0.18**
Ketamine 35 mg/kg, i.p.	84 ± 6**	7.37 ± 0.04	81.67 ± 0.11**	34.38 ± 0.19*
Pentobarbital 50 mg/kg, i.p.	80 ± 6**	7.36 ± 0.05	81.89 ± 0.21**	34.78 ± 0.12**
Equithesin 3 mL/kg, i.p.	94 ± 8**[a]	7.37 ± 0.06	81.15 ± 0.18*[a]	34.35 ± 0.16*[a]
Urethane# 1 g/kg, i.p.	90 ± 5**[a]	7.36 ± 0.08	81.19 ± 0.14*[a]	34.32 ± 0.21*[a]

[a]$P < 0.05$ from ether.

Values are mean ± SD from six to eight rats at each point. ANOVA followed by Dunnet's test for multiple group comparison. * = $P < 0.05$; ** $P < 0.01$ from respective controls.

(Table 1). The most marked reduction in the MABP after CHI was seen under ether anesthesia followed by pentobarbital and ketamine. The least reduction on CHI induced MABP was observed in animals that were anesthetized with ketamine and urethane.

Arterial pH from control group under ether, pentobarbital and urethane showed more acidic pH after CHI as compared to ketamine and Equithesin (Table 2). However, these differences were not significant.

Significant increase in the PaO_2 was observed after CHI in groups that received ether followed by pentobarbital and ketamine. Urethane and Equithesin also showed a mild but significant increase in PaO_2 after CHI as compared to same anesthetic exposure in normal animals (Table 1). Likewise significant increase in $PaCO_2$ was also seen after CHI in ether and pentobarbital anesthetized animals. Whereas, Ketamine, Equithesin and urethane anesthetized rats showed only a mild but significant increase in $PaCO_2$ after CHI as compared to same anesthetics in normal animals (Table 2).

3.2 Functional outcome in concussive head injury

Normal animals anesthetized with different anesthetics for 5 h were subjected to several sensorimotor and cognitive function tests after regaining consciousness shows no significant differences on inclined plane angle test or placement error on a mesh grid (Table 2). However, significant differences between time of stay on the Rota Rod treadmill at 16 rpm during a 120 sex session is seen in normal animals after different anesthetics (Table 2). The most remarkable reduction of time on staying at Rota Rod treadmill in normal animals was seen in ether anesthetized rats as compared to all other anesthetics (Table 2). Equithesin has the best effects on staying at Rota Rod treadmill (100%) followed by ketamine and pentobarbital (Table 2).

However, when CHI was inflicted under these anesthetics significant changes in functional outcome was seen (Table 2). CHI under Equithesin showed the highest scores on all parameters of functional outcome such as stay time on Rota Rod, walking on inclined plane angel, nr of steps taken on a mesh grid and least foot placement errors (Table 2). These good scores of Equithesin were followed by pentobarbital and ketamine in CHI on several functional outcome measures. Ether anesthetized CHI group showed the worst performance on all the functional outcome parameters (see Table 2).

Table 2 Effect of anesthetics on functional outcome, serotonin levels and cortical brain pathology.

	Rota Rod	Inclined plane°	Grid walking		Plasma 5-HT	Brain 5-HT (μg/g)		Nerve cell damage nr.	
Type of expt.	**(S)**	**Angle°**	**Nr of steps**	**Error %**	**(μg/mL)**	**Right**	**Left**	**Right**	**Left**
A. Control									
Ether inhalation	110±2	58±2	38±3	4±1	0.48±0.06*	0.98±0.08*	1.04±0.07*	2±1	2±2
Ketamine 35 mg/kg, i.p.	118±1*	58±1	40±2	2±1	0.40±0.03*	0.85±0.06*	0.87±0.09*[a]	nil	nil
Pentobarbital 50 mg/kg, i.p.	116±3*	57±2	38±2	1±1	0.38±0.04*	0.82±0.07*	0.80±0.06*[a]	nil	nil
Equithesin 3 mL/kg, i.p.	120±1*	60±1	40±4	1±1	0.34±0.04*	0.78±0.12*	0.74±0.09*[a]	nil	nil
Urethane#1 g/kg, i.p.	ND	ND	ND	ND	0.26±0.07[a]	0.67±0.21[a]	0.69±0.13[a]	nil	nil
B. Concussive head injury (CHI)									
Ether inhalation	50±3*	22±2*	10±2*	48±2*	1.56±0.06*	2.89±0.21*	3.59±0.10*[b]	146±12*	164±18*[b]
Ketamine 35 mg/kg, i.p.	58±3*[a]	26±3*[a]	14±2*[a]	44±4*[a]	1.28±0.06*[a]	2.56±0.21*[a]	3.36±0 .13*[a,b]	118±8*[a]	143±7*[a,b]
Pentobarbital 50 mg/kg, i.p.	62±6*[a]	30±3*[a]	18±4*[a]	38±3*[a]	1.14±0.12*[a]	2.46±0.10*[a]	3.13±0.14*[a,b]	112±10*[a]	138±12*[a,b]
Equithesin 3 mL/kg, i.p.	70±3*[a]	35±3*[a]	22±2*[a]	32±3*[a]	0.98±0.08*[a]	2.34±0.21*[a]	2.86±0.22*[a,b]	84±4*[a]	96±8*[a,b]
Urethane# 1 g/kg, i.p.	ND	ND	ND	ND	0.84±0.03[a]	1.78±0.10[a]	2.08±0.13[a,b]	76±8[a]	89±7[a,b]

[a] $P<0.05$ from ether.
[b] $P<0.05$ from right half.
Values are mean ± SD from six to eight rats at each point. ANOVA followed by Dunnet's test for multiple group comparison. * = $P<0.05$ from respective control. Neuronal damage was evaluated using Chi-Square test.
Irreversible anesthesia thus functional outcome could not be measured in these animals.

3.3 Blood-brain barrier in concussive head injury

We measured BBB breakdown using Evans blue and radioiodine tracers in normal and CHI rats under different anesthetics (Table 3; Fig. 2).

Control normal animals anesthetized with different anesthetics showed minor but significant changes in the tracer accumulation in the right or left half in control group without any trauma (Table 3). Thus, in comparison to urethane anesthetized rats normal BBB function showed greater accumulation of Evans blue or radioiodine within the brains in rats with other anesthetics. Whereas as compared to ether anesthetized animals accumulation of these tracers ware significantly lower in the brains of left and right halves in rats with all other anesthetics (Table 3). However, in normal animals no significant differences between left or right halves were seen with regard to either Evans blue or radioiodine tracers accumulation (Table 3). In general, urethane showed the least accumulation of Evans blue and radioiodine in the brain followed by Equithesin, pentobarbital, ketamine and ether (Table 3).

When CHI was inflicted under these different anesthetics, extravasation of these tracers was significantly different in the right injured and left contralateral hemisphere. Since our CHI represents counter-coup injury, the left uninjured side exhibited greater extravasation of these tracers than the inurned right hemisphere. Interestingly, this increase in BBB to Evans blue and radioiodine in the left or right hemisphere also varied significantly among different anesthetics (See Table 3).

CHI performed under urethane anesthesia, the mean values of Evans blue and radioiodine leakage in the right and left halves of the brain was the lowest among all other anesthetized agents after CHI (Table 3). In general ether anesthetized group after CHI showed the highest increase in tracer extravasation in both the right and left hemispheres. Interestingly, in ether anesthetized group the differences between left and right hemisphere with regard to tracer extravasation did not differ although a higher tendency of BBB leakage was still evident in the left side as compared to the injured right half (Table 3). The magnitude and intensity of the BBB breakdown in the left and right side in CHI under urethane, Equithesin exhibited a significantly mild increase from the urethane group that was followed by pentobarbital and ketamine for both Evans blue and radioiodine (see Table 3).

When the whole brain Evans blue was measured comprising both the right and left hemispheres after CHI slight changes in this pattern is visible (see Fig. 2).

Table 3 Effect of anesthetics on concussive head injury induced cortical brain pathophysiology in rats.

Type of Expt.	EBA mg %		[131]Iodine %		Brain water %		CBF (mL/g/min)	
	Right	Left	Right	Left	Right	Left	Right	Left
A. Control								
Ether inhalation	0.29 ± 0.06*	0.30 ± 0.07*	0.44 ± 0.09*	0.43 ± 0.06*	75.34 ± 0.32*	75.67 ± 0.21*	1.23 ± 0.08	1.26 ± 0.05
Pentobarbital 50 mg/kg, i.p.	0.24 ± 0.05*[a]	0.22 ± 0.04*[a]	0.30 ± 0.04*[a]	0.33 ± 0.05*[a]	74.56 ± 0.23[a]	74.51 ± 0.21[a]	1.20 ± 0.04	1.18 ± 0.06
Ketamine 35 mg/kg, i.p.	0.26 ± 0.04*[a]	0.28 ± 0.08*	0.34 ± 0.08*[a]	0.36 ± 0.04*[a]	74.78 ± 0.24[a]	74.77 ± 0.32[a]	1.18 ± 0.05	1.16 ± 0.03
Equithesin 3 mL/kg, i.p.	0.22 ± 0.06*[a]	0.24 ± 0.08*[a]	0.32 ± 0.05*[a]	0.34 ± 0.08*[a]	74.67 ± 0.45[a]	74.48 ± 0.43[a]	1.18 ± 0.04	1.16 ± 0.06
Urethane 1 g/kg, i.p.	0.18 ± 0.02*[a]	0.20 ± 0.04*[a]	0.28 ± 0.06*[a]	0.30 ± 0.04*[a]	74.76 ± 0.13[a]	74.78 ± 0.23[a]	1.20 ± 0.08	1.21 ± 0.06
B. Concussive head injury (CHI)								
Ether inhalation	1.89 ± 0.23*	2.14 ± 0.12*[a]	3.23 ± 0.21*	3.67 ± 0.34*[a]	78.47 ± 0.34*	79.48 ± 0.43*	0.70 ± 0.08*	0.60 ± 0.06*[a]
Ketamine 35 mg/kg, i.p.	1.75 ± 0.08*[a]	1.94 ± 0.07*[a,b]	2.21 ± 0.04*[a,b]	2.96 ± 0.10*[a,b]	78.04 ± 0.12*[a]	78.98 ± 0.23*[a,b]	0.78 ± 0.04*[a]	0.70 ± 0.03*[a,b]
Pentobarbital 50 mg/kg, i.p.	1.67 ± 0.11*[a]	2.01 ± 0.10*[a,b]	2.32 ± 0.06*[a]	2.84 ± 0.06*[a,b]	77.78 ± 0.10*[a]	78.66 ± 0.33*[a,b]	0.82 ± 0.04*[a]	0.76 ± 0.04*[a,b]
Equithesin 3 mL/kg, i.p.	1.45 ± 0.08*[a]	1.88 ± 0.06*[a,b]	2.14 ± 0.08*[a]	2.67 ± 0.10*[a,b]	77.21 ± 0.32*[a]	78.34 ± 0.21*[a,b]	0.88 ± 0.08*[a]	0.80 ± 0.04*[a,b]
Urethane 1 g/kg, i.p.	1.24 ± 0.06*[a]	1.44 ± 0.07*[a,b]	1.89 ± 0.08*[a]	2.14 ± 0.08*[a,b]	76.98 ± 0.21[a]	77.78 ± 0.43[a,b]	0.96 ± 0.08*[a]	0.90 ± 0.04*[a,b]

[a] $P < 0.05$ from ether.
[b] $P < 0.05$ from right half.
Values are mean ± SD from six to eight rats at each point. * $P < 0.05$ from urethane.
ANOVA followed by Dunnet's test for multiple group comparison.
Right = injured with 0.224 N impact over skull bone; Left = uninjured on skull, but counter coup concussion is prominent; CHI = concussive head injury; Brain Water = cortical water content; CBF = cortical blood flow; EBA = Evans blue albumin = Cortical extravasation; Radioiodine = cortical extravasation.

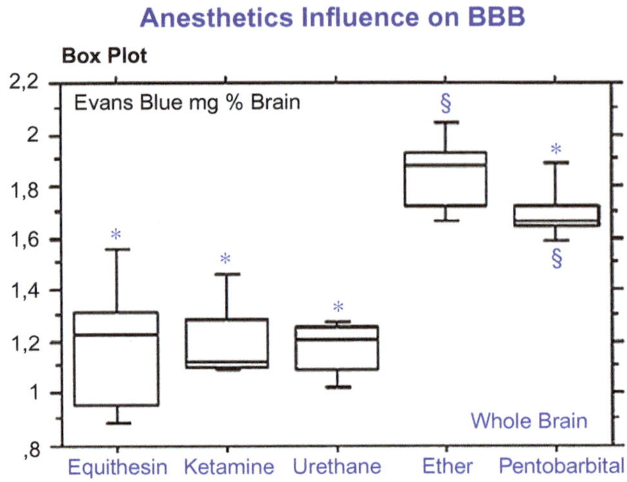

Fig. 2 Shows blood-brain barrier (BBB) disruption to Evans blue (mg %) in whole brain after concussive head injury (CHI) delivered on the right parietal skull in anesthetized rats using different anesthetics. Animals were allowed to survive 5 h after injury. Leakage of Evans blue was measured in whole brain using sensitive spectrophotometric determination using standard protocol. For details see text. Each box represents data from six to eight animals. * <0.05 from Ether anesthetized group, § $P<0.05$ from urethane-anesthetized group. ANOVA followed by Dunnett's test for multiple group comparison from one control.

3.4 Brain edema in concussive head injury

Brain edema was measured in CHI group using water content of the right injured and left uninjured half under different anesthetics and compared them with normal animals under identical anesthesia (See Table 3).

In normal animals, water content of the left or right hemisphere did not differ significantly under any anesthetics. However, there are minor but significant differences in brain water contents under different anesthetics over 5 h period (Table 3).

In normal animals anesthetized with urethane exhibited mean 74.76% water in the right half and 74.78% brain water in the left half. On the other hand, ether anesthetized animals showed significantly higher brain water in both left (75.76%) and in right (75.34%) half. The brain water values in right or left halves were not different from urethane significantly in Equithesin, ketamine or pentobarbital anesthesia in normal animals. Thus, brain water under all these anesthetics were significantly lower than that seen under ether anesthesia in normal animals (Table 3).

Interestingly, after CHI under urethane anesthesia the rise of brain water was also the least in both left and right hemispheres as compared to all other anesthetics. In CHI, left contralateral half exhibited significantly higher accumulation of brain water content than the right injured half under urethane anesthesia like all other anesthetics (Table 3). The highest brain water increase following CHI was seen under ether anesthesia where the differences between right and left halves ware not found significant although the left side showed a greater accumulation of brain water (Table 3). The whole brain appears to be swollen after CHI under ether anesthesia. The brain water after urethane showed greater increase in Equithesin anesthetized injured rats followed by pentobarbital and ketamine (Table 3). In these CHI groups left half showed significantly higher accumulation in brain water that the right injured side as seen under urethane anesthesia (Table 3). However, this increase in brain water content in right half under Equithesin, pentobarbital and ketamine are significantly lower than the left half of brain in CHI (see Table 3).

When the water content was measured in whole brain comprising left and right halves, some minor changes in the pattern under different anesthetics were observed (See Fig. 3).

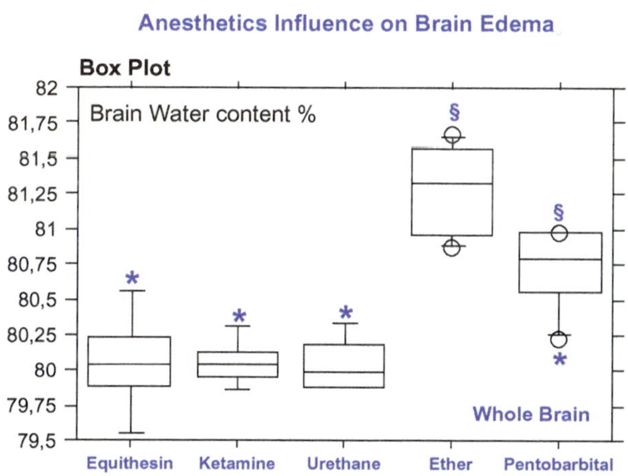

Fig. 3 Shows brain water content in whole brain after concussive head injury (CHI) delivered on the right parietal skull in anesthetized rats using different anesthetics. Animals were allowed to survive 5 h after injury. Increased brain water content representing brain edema significantly differs after CHI under various anesthetics used. For details see text. Each box represents data from six to eight animals. * <0.05 from Ether anesthetized group, § $P<0.05$ from urethane-anesthetized group. ANOVA followed by Dunnett's test for multiple group comparison from one control.

3.5 Cerebral blood flow in concussive head injury

We measured cerebral blood flow (CBF) in right and left cerebral cortex using carbonized microsphere labeled with radioiodine in in CHI under different anesthetics and compared the results with normal animals under the same anesthesia. The results are shown in Table 3.

Different anesthetics in normal animals showed minor differences in cortical CBF between the left or right halves and among different normal animal groups under various anesthetics (Table 3). However, after CHI under these anesthetics induced remarkable significant reductions in the cortical CBF in both the right and the left halves. This reduction in the CBF was the most prominent in the left uninjured half as compared to the right traumatized half (Table 3).

With regard to CHI under urethane aesthesia, the reduction in CBF is the minimum in both right and left halves after 5 h trauma. This reduction in the cortical CBF is significantly lower in the left half as compared to the right injured half (Table 3). In the CHI group under ether anesthesia greatest reduction in the CBF was seen in both halves. The left half showing a significantly higher reduction in CBF after CHI in ether anesthesia than in the right injured half (Table 3). The CBF reductions in CHI were significantly lower in ketamine-anesthetized rats followed by pentobarbital and Equithesin anesthesia in both halves. However, left uninjured half in all these anesthetic groups showed significantly lower values than the right traumatized half (Table 3).

Measurement of CBF in the whole brain comprising right and left halves showed some minor modification in under different anesthetics (see Fig. 4).

3.6 Serotonin levels in concussive head injury

We measured plasma and brain serotonin levels using a highly sensitive and specific spectrophotofluorometric method to determine the amine in the plasma and brain in normal and CHI groups under different anesthesia (Table 2).

Our observations show that serotonin level varied significantly in normal animals under different anesthetics (See Table 2). The serotonin levels in the plasma and in the right and left cerebral cortical tissues were found at the lowest levels under urethane anesthetized normal animals. Whereas, the highest levels of the amine was seen in plasma and brains of ether anesthetized control rats (Table 2). As compared to ether anesthetized normal animals, serotonin levels were significantly lower in the plasma and brains

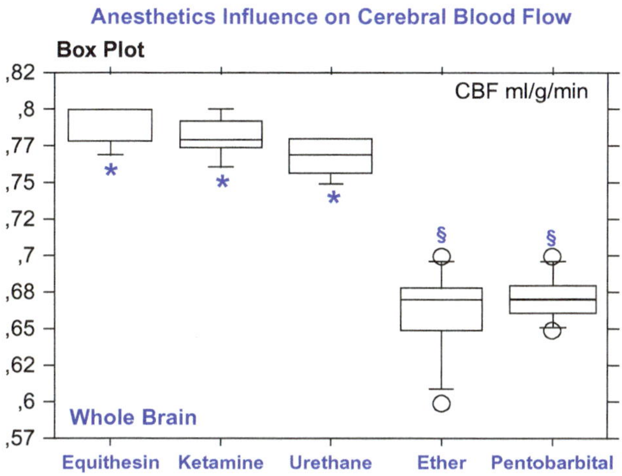

Fig. 4 Shows cerebral blood flow (CBF) in whole brain after concussive head injury (CHI) delivered on the right parietal skull in anesthetized rats using different anesthetics. Animals were allowed to survive 5 h after injury. CBF was measured using microsphere techniques in the whole brain that significantly differs after CHI under various anesthetics used. For details see text. Each box represents data from six to eight animals. * <0.05 from Ether anesthetized group, § $P < 0.05$ from urethane-anesthetized group. ANOVA followed by Dunnett's test for multiple group comparison from one control.

in ketamine-anesthetized rats followed by pentobarbital and Equithesin (Table 3). In normal animals differences between right and left cortical tissues serotonin concentration was not significantly different in any groups under any anesthetics (see Table 2).

However, CHI inflicted in animals under various anesthetics significantly enhanced the plasma and brain serotonin levels in all groups. After CHI, the increase in serotonin levels in the left uninjured half was higher than the right traumatized half in all animals irrespective of the anesthetics used (Table 2).

Under urethane anesthesia CHI resulted in serotonin elevation in the right half of mean value 1.78 μg/g the left half showed 2.08 μg/g content of the amine. The plasma values also rose in this CHI group to mean 0.84 μg/mL (Table 2). In ether anesthetized rats with identical CHI resulted in the highest increase in plasma by 1.56 μg/mL and the injured side of the brain exhibited 2.89-μg/g tissue levels. The left side showed a significant higher level of the amine 3.59-μg/g content after CHI under ether anesthesia (Table 2). Equithesin anesthetized CHI rats showed only a mild but significant elevation of plasma and brain serotonin levels followed by pentobarbital and ketamine anesthetized rats in CHI (Table 2). Interestingly in all these groups serotonin content in the left half was significantly higher that the right half after CHI (Table 2).

3.6.1 Correlation between plasma serotonin and brain edema formation

Serotonin is one of the key mediators of the BBB and vasogenic brain edema formation (Baethmann et al., 1988; Wahl, Unterberg, Baethmann, & Schilling, 1988, 1990). Thus, we used regression analysis to establish the relationship between plasma serotonin and brain edema formation in normal and CHI group irrespective of anesthetics used. Our analysis showed a very tight correlation between plasma serotonin and brain edema formation in CHI groups under various anesthetics. The correlation coefficient r2 was 0.945 (see Fig. 5). This suggests that anesthetic induce changes in the plasma serotonin could be one of the key players in determining brain edema in CHI.

3.7 Nerve cell damage in concussive head injury

Traumatic injuries to the CNS such as CHI could lead to serious complications and life threatening challenge because of severe damages to the underlying brain tissue (Blyth & Bazarian, 2010; Ptyushkin, Vidmar, Burger, & Marincek, 2012; Wood, 2008). Thus, we compared neuronal damages in the left and right halves of traumatized brains under different anesthetics (see Table 2).

We used standard histopathological techniques to study neuronal damages on paraffin sections stained with H&E (See Fig. 6).

Normal animals without trauma when anesthetized with different anesthetics for 5 h and brain tissues were examined for neuronal damages, we cannot find any significant neuronal distortion or damages in the brain of rats anesthetized by any anesthetics (Table 2).

However, after CHI the number of damaged or distorted neurons significantly increased in bath the right injured and the left contralateral critical tissues after 5 h period irrespective of anesthesia used (Table 2). In general, the left half showed significantly higher neuronal damages than the right injured half in all CHI groups under different anesthetics (Table 2). However, under urethane anesthesia the number of injured neurons in both the right and left side is the lowest. The highest number of injured neurons in both halves of brain following CHI is seen in animals under ether anesthesia followed by ketamine and pentobarbital. Equithesin anesthetized rats after

Anesthetics influence on Serotonin

Regression Coefficients
plasma 5–HT vs. Brain edema

	Coefficient	Std. Error	Std. Coeff.	t-Value	P-Value
Intercept	–31,863	1,703	–31,863	–18,709	<,0001
Brain edema	,425	,022	,972	19,400	<,0001

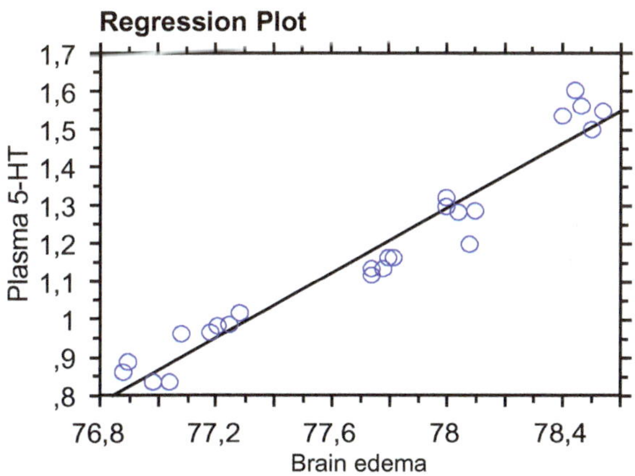

Y = –31,863 + ,425* X; R^2 = ,945

Fig. 5 Shows regression analysis between plasma serotonin (5-HT) level (μg/mL) with brain water content (%) as measure of edema in control and CHI rats under all anesthetics. A tight correlation (r2 = 0.945, $P < 0.001$) between plasma serotonin and brain water content is seen in CHI irrespective of anesthetics used. For details see text.

CHI showed only a mild but significant increase in the number of damaged neurons in the brain. The left side always showed greater number of damaged neurons than the right half (Table 2).

Morphological examination showed several damaged and distorted neurons in the neuropil of right and left halves in the brain following CHI in different anesthetic groups. Some neurons were swollen and some were shrunken (Fig. 6). These damaged neurons are located largely within the edematous neuropil showing sponginess and general expansion (Fig. 6). These changes were most marked in CHI under ether anesthesia followed by ketamine, pentobarbital, Equithesin and urethane (Fig. 6; Table 2).

A representative example of brains damage in the parietal cortex of right and left under urethane, Equithesin, pentobarbital and ether are shown in Fig. 6. From these images severe damage to neurons is evident in ether-anesthetized rats after CHI as compared to other anesthetics. CHI induced

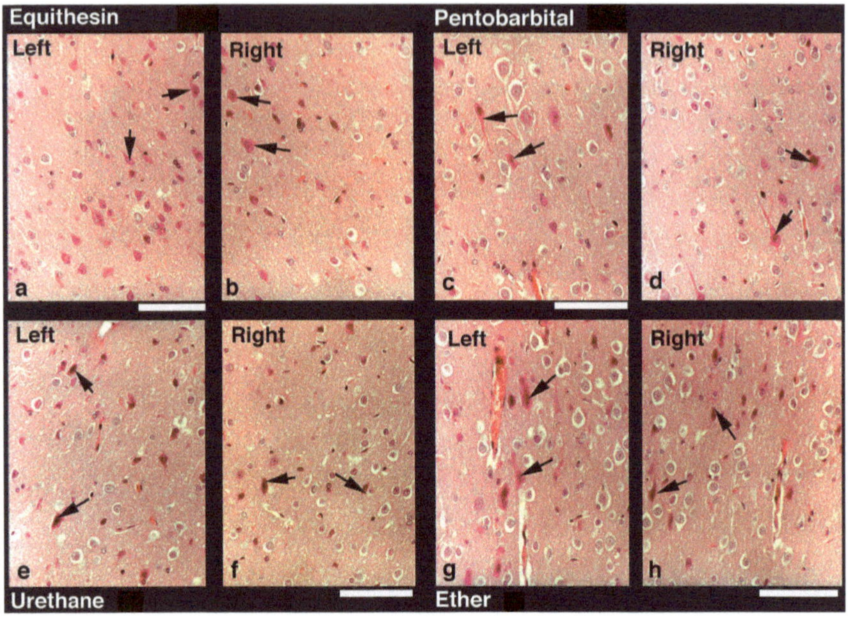

Fig. 6 Shows neuronal damages in right injured and left contralateral side following 5 h CHI in the rat parietal cerebral cortex under Equithesin (a, b), Pentobarbital (c, d), Urethane (e, f) and Ether (g, h) anesthesia. Profound neuronal damages with perineuronal edema are seen that is most frequent under Ether (g, h) and Pentobarbital (c, d) anesthesia (arrows). Under Ether anesthesia both right and left side (g, h) is almost equally damaged. Whereas Equithesin (a, b) and Urethane (e, f) showed neuronal damages with dark and distorted neurons spread over the wide areas of the cortex (arrows) however, perineuronal edema is not distinct as compared to Ether and Pentobarbital. Hematoxylin and Eosin (h&e) staining on 3-μm thick paraffin sections from rat parietal cerebral cortex. Scale Bars = 35 μm. For details see text.

minimal damages to neurons are seen under urethane anesthesia followed by Equithesin. These observations strongly support the idea that anesthetics could influence cellular damages in CHI.

4. Possible mechanisms and functional significance

The salient new findings from our investigations clearly show that the pathophysiology of CHI varies under different anesthesia. The outcome of CHI with regard to the BBB breakdown, edema formation and alterations in the CBF are much lower in urethane-anesthetized rats (Field & Lang, 1988; Maggi & Meli, 1986a, 1986b, 1986c; Pagliardini et al., 2013a; Pagliardini,

Funk, & Dickson, 2013b). On the other hand highest degree of these changes after CHI is seen under ether anesthesia (Bourne, 1946; Bovill, 2008; Defalque & Wright, 2008; England & Jones, 1992; Finer, 1965; Ghosh & Banerjee, 1965; Greuter et al., 2017; No Author Listed, 1852). Ketamine and Equithesin (Deacon & Rawlins, 1996; Laaksonen et al., 2018; Li et al., 2018; Mueller & Hunt, 1998; Muir, 2010; Yan & Jiang, 2014) induced these changes are very close to that of urethane while pentobarbital shows somewhat greater magnitude of these changes that urethane.

Our observations further show that these changes in the BBB, brain edema formation and CBF exhibited a close parallelism between plasma and brain serotonin levels in CHI under different anesthetics. This suggests that peripheral and central serotoninergic system is one of the important regulators playing key roles in the pathophysiology of CHI under different anesthetics (Abe, Shimada, Okada, & Kibayashi, 2016; Adler, Kuriyama, & Maynert, 1965; Buxbaum, Yarbrough, & Carter, 1973; Jin et al., 2016; Kosari-Nasab et al., 2019; Li et al., 2019; Liepert, 2016; Moore, Wong, & Heller, 1965).

Head injury either caused by blunt forces, fall, sorts injury such as rugby, jumping from parachute leads to various symptoms from sleep disturbances, mental confusion, delirium, headache and concussion (Armistead-Jehle, Soble, Cooper, & Belanger, 2017; Helmick et al., 2015; Rona, Jones, Jones, Fear, & Wessely, 2019; Sabourin, Holland, Mau, Gandhi, & Prestigiacomo, 2016). These changes could induce slowly developing neurodegeneration in the CNS leading to Alzheimer's disease and Parkinson's disease with lifetime disability (Pattinson et al., 2019; Pieper et al., 2019; Stein, 2019).

Most head injury cases come from military during combat operations. Bullet wounds, piercing injuries, missile or blast injuries lead to hemorrhage, tissue loss, severing of neuronal network all could result in permanent lifetime disability (Hicks, James, Spitz, & Ponsford, 2019; Nicoletti et al., 2017; Rudelli, Strom, Welch, & Ambler, 1982; Takizawa, Gemmell, Kenworthy, & Speyer, 2016). Thus, studies regarding pathophysiology of head injuries require further exploration with regard to expand our understanding towards basic mechanisms or new therapeutic advances.

Our observations showed that CHI performed under different anesthetic agents have significantly different outcome on pathophysiology and functional parameters. This suggests that anesthetic influence both the functional outcome and brain pathology following CHI. Thus, it is quite likely that different anesthetics will influence therapeutic potentiates of drug therapies in clinics. However, this is a new feature that requires detailed investigations in CHI.

Anesthetics induce profound stress at the cellular level and induce gene expression in the hippocampus within short period of time (Hen, Wei, Chen, Huang, & Chen, 2018; Hoetzel & Schmidt, 2010; Mantz & Dahmani, 2008; Pekny, Andersson, Wilhelmsson, Pekna, & Pekny, 2014; Perouansky, 2008). General anesthesia is well known to induce neurogenesis and increased amyloid beta peptide levels in the brain (Berger et al., 2016; Jiang & Jiang, 2015; Zhang et al., 2017; Zhou & Bickler, 2017). Isoflurane anesthesia is able to induce >20 genes responsible for cellular stress, neural plasticity response and hippocampal cell injury within 10 min of general anesthesia that lasted for >4 days in mice (Hen et al., 2018; Pekny et al., 2014). These observations suggest that anesthetics could induce marked alterations in cellular and molecular mechanisms of the CNS.

Anesthetics induce stress could stimulate serotonin system that is very sensitive to hypothalamic-pituitary-adrenal axis (Etsten & Shimosato, 1966; Hwang, Hyun, Kim, Kim, & Shin, 2017; Korneyev, Costa, & Guidotti, 1993; Li, Dong, Tan, Qian, & Jin, 2019). Increased plasma level of serotonin in ether anesthesia as compared to urethane and other anesthetics in normal animals support this idea (BARCHAS & DX, 1963; Natarajan, Northrop, & Yamamoto, 2015; Overgaard, Lieblich, Richardson, Galea, & Frokjaer, 2018; Puglisi-Allegra & Andolina, 2015; RM, 1960). Both peripheral and central serotonergic system is affected by stress (Chaouloff, 1993; Fuller, 1981; Murnane, 2019) This is evident from the findings that stress induced by immobilization; heat or forced swimming caused serotonin release in the plasma and brain (Lafuente et al., 2018; Sharma et al., 1992; Sharma & Cervós-Navarro, 1990; Sharma, Cervós-Navarro, & Dey, 1991b; Sharma & Dey, 1981, 1986a, 1986b, 1988; Sharma & Sharma, 2007; Sharma, Westman, Navarro, Dey, & Nyberg, 1995). There was a close relationship between increased circulating level of serotonin and breakdown of the BBB to Evans blue and radioiodine tracers (see Sharma, 2004a, 2004b). This suggests that elevated serotonin levels in plasma are crucial for BBB breakdown to protein tracers.

Further support to this hypothesis comes from drug treatments. Thus, inhibition of serotonin synthesis with *p*-chlorophenylalanine (p-CPA) significantly reduced the BBB breakdown and serotonin levels in stress caused by immobilization, heat or forced swimming (Sharma, 2004a, 2004b). Whether serotonin has a direct effects on cerebral endothelial cells in opening of the BBB to protein tracers is further confirmed when exogenous serotonin was administered into the circulation leading to significant increase in the BBB permeability to Evans blue and radioiodine tracers (Sharma et al., 1990, 1991b, 1992, 1993, 1995; Sharma & Cervós-Navarro, 1990; Sharma & Dey, 1981, 1986a, 1986b, 1988; Sharma & Sharma, 2007). This suggests that elevation of serotonin in the plasma contributes to the BBB breakdown.

Topical application of serotonin on the pial surfaces as well as intracarotid infusion of serotonin induces BBB breakdown, alterations in CBF and vasogenic edema formation (Mohanty, Dey, & Ray, 1979; Sharma, 2004a, 2004b; Sharma, Dey, & Olsson, 1989; Sharma, Olsson, & Dey, 1989). Likewise human brain contusion and edematous areas exhibited increased serotonin content confirming the role of 5-HT in cerebral edema (Mohanty et al., 1979; Mohanty, Dey, Sen, & Ray, 1978; Mohanty & Mazumdar, 1978; Sharma et al., 2007b). After CHI microvascular permeability disturbances release blood platelets containing serotonin into the circulatory system that is further compounded by the stress of anesthetics resulting in massive accumulation of the amine in blood and brain causing edema formation and CBF disturbances (Dey & Sharma, 1984; Sharma & Dey, 1986a, 1986b). This hypothesis is supported from the fact that infusion of serotonin antibodies into the cerebral ventricles after CHI attenuated edema formation and cell injury (see Sharma et al., 2007b; Sharma, Winkler, Stålberg, Mohanty, & Westman, 2000; Vannemreddy et al., 2006b).

That serotonin plays key roles in anesthetics induced changes in the BBB breakdown, CBF and edema formation is further supported by the findings that showed a close correlation between plasma serotonin and brain edema formation. BBB breakdown will allow extravasation of serum proteins into the cerebral compartment leading to vasogenic edema formation (Fig. 5). Greater increase in the BBB disruption to protein tracers Evans blue and radioiodine in the contralateral side associated with higher brain edema formation are in line with this idea (Sharma et al., 2000, 2007b; Vannemreddy et al., 2006b). Evans blue and radioiodine when introduced into the systemic circulation will binds to several molecules of serum proteins (Sharma & Dey, 1986b, 1988). Thus, when leakage of these tracers occurs that represents extravasation of tracer-protein complex. In this study, Evans blue leakage that is normally bound to 12 molecules of albumin to one molecule of the dye making is dye-protein complex (MW 68 kDa) showed profound extravasation in the edematous areas showing cell injury (Table 2).

The level of anesthetic stress resulting in varied degree of edema formation, BBB breakdown, CBF alterations and circulating serotonin levels further compounds these changes induced by trauma. Several lines of evidences show that use of anesthetics also influence surgical outcome in several cancer cases and other diseases. It remains to be seen what are the key factors that influence anesthetics induced changes in various outcomes after surgery or related procedures (Budniewski, 1975; Ivankovich et al., 1976; Komatsu, Nogaya, & Ogli, 1994; Pettersson & Akerman, 1979; Wagner, Frid, Evtukhin, Barchuk, & Shulga, 1981).

It appears that anesthetics could also affect differently with regard to their binding to NMDA receptor complex, gene expression, GABA receptor antagonisms and other molecular mechanisms to induce their actions (Brohan & Goudra, 2017; Komatsu et al., 1994; Muir, 2010; Olsen & Li, 2011; Petrenko et al., 2014; Saari et al., 2011; Sharma & Sjöquist, 2002). All these factors contribute to CHI induced pathophysiology as seen in our investigation. Our results further show a close parallelism between functions outcome and brain pathology. This suggests that brain pathology caused by CHI is somehow related to functional disturbances. However, further studies are needed to clarify these points.

5. Conclusion and future perspectives

In conclusion, our investigations show that anesthetics influence pathophysiology of CHI. Ether anesthetic induces profound enhanced brain pathology whereas urethane and Equithesin exhibited significant but minimal changes in brain pathology as compared to other anesthetics such as ketamine and pentobarbital. Since urethane is an irreversible anesthesia, its use is limited for experimental procedures only. For clinical use, Equithesin and ketamine appears to be ideal for minimizing CHI induced brain pathology. Our results also showed that the relationship between CHI induced brain pathology with anesthetics appear to be mediated via agnostics induced stress at the general or cellular level. Stress induced by anesthesia results in serotonin upregulation in plasma and brain that shows a strong correlation between plasma serotonin and development of brain edema. However, anesthetics such propofol is known to induce heat shock protein regulation in various brain areas, it remains to be seen whether these anesthetics also affect molecular mechanisms lf different genes involved in brain pathology, such as nitric oxide synthase (NOS), hemeoxygenase (HO), tumor necrosis factor alpha (TNF-α) and other harmful cytokines, dynorphin peptide and other substances involved in neurotoxicity and/or neuroprotection.

Another important aspect of anesthetic influence that is the most relevant in clinics is to find our whether these anesthetics could also influence neuroprotective ability of different compounds used to treat neurodegenerative diseases of neurological cases such as stroke, brain tumor, traumatic injuries to the brain or spinal cord, a feature that is currently being investigated in our laboratory.

Acknowledgments

Supported by grants from the Swedish Medical Research Council (Nr 2710-HSS), Göran Gustafsson Foundation, Stockholm, Sweden (H.S.S.), Astra Zeneca, Mölndal, Sweden (H.S.S./A.S.), the National Institutes of Health (R01 AG028679); Laerdal Foundation of Acute Medicine, Norway (H.S.S., L.W.); The University Grants Commission, New Delhi, India (H.S.S./A.S.), Ministry of Science & Technology, People Republic of China; Ministry of Science & Technology, Govt. of India (H.S.S./A.S.), Indian Medical Research Council, New Delhi, India (H.S.S./A.S.) and India-EU Co-operation Program (A.S./H.S.S.) & Society for Neuroprotection and Neuroplasticity (SSNN), Romania. Expert technical assistance on morphology by Kerstin Flink and Ingmarie Olsson (Uppsala University) is highly appreciated with thanks. We thank Suraj Sharma, Uppsala, Sweden for computer and graphic support.

References

Abe, K., Shimada, R., Okada, Y., & Kibayashi, K. (2016). Traumatic brain injury decreases serotonin transporter expression in the rat cerebrum. *Neurological Research*, *38*(4), 358–363. https://doi.org/10.1080/01616412.2015.1110402. Epub 2016 Apr 15.

Adler, M. W., Kuriyama, K., & Maynert, E. W. (1965). Norepinephrine and 5-hydroxytryptamine in rat brain stem following cortical lesions. *Life Sciences (1962)*, *4*, 141–144. No abstract available.

Alam, A., Suen, K. C., Hana, Z., Sanders, R. D., Maze, M., & Ma, D. (2017). Neuroprotection and neurotoxicity in the developing brain: An update on the effects of dexmedetomidine and xenon. *Neurotoxicology and Teratology*, *60*, 102–116. https://doi.org/10.1016/j.ntt.2017.01.001. Epub 2017 Jan 6. Review.

Alkire, M. T., Hudetz, A. G., & Tononi, G. (2008). Consciousness and anesthesia. *Science*, *322*(5903), 876–880. https://doi.org/10.1126/science.1149213.

Andropoulos, D. B. (2018). Effect of anesthesia on the developing brain: Infant and fetus. *Fetal Diagnosis and Therapy*, *43*(1), 1–11. https://doi.org/10.1159/000475928. Epub 2017 Jun 7.

Armistead-Jehle, P., Soble, J. R., Cooper, D. B., & Belanger, H. G. (2017). Unique aspects of traumatic brain injury in military and veteran populations. *Physical Medicine and Rehabilitation Clinics of North America*, *28*(2), 323–337. https://doi.org/10.1016/j.pmr.2016.12.008.

Baethmann, A., Maier-Hauff, K., Kempski, O., Unterberg, A., Wahl, M., & Schürer, L. (1988). Mediators of brain edema and secondary brain damage. *Critical Care Medicine*, *16*(10), 972–978. Review.

BARCHAS, J. D., & DX, F. (1963). Brain amines: Response to physiological stress. *Biochemical Pharmacology*, *12*, 1232–1235. No abstract available.

Belopavlovic, M., & Buchthal, A. (1980). Barbiturate therapy in cerebral ischaemia. *Anaesthesia, 35*(3), 235–245. Review.

Berger, M., Nadler, J. W., Friedman, A., McDonagh, D. L., Bennett, E. R., Cooter, M., et al. (2016). MAD-PIA trial team. The effect of propofol versus isoflurane anesthesia on human cerebrospinal fluid markers of Alzheimer's disease: Results of a randomized trial. *Journal of Alzheimer's Disease, 52*(4), 1299–1310. https://doi.org/10.3233/JAD-151190.

Bilotta, F., Evered, L. A., & Gruenbaum, S. E. (2017). Neurotoxicity of anesthetic drugs: An update. *Current Opinion in Anaesthesiology, 30*(4), 452–457. https://doi.org/10.1097/ACO.0000000000000482. Review.

Blyth, B. J., & Bazarian, J. J. (2010). Traumatic alterations in consciousness: Traumatic brain injury. *Emergency Medicine Clinics of North America, 28*(3), 571–594. https://doi.org/10.1016/j.emc.2010.03.003. Review.

Bourne, W. (1946). Pure ether and impurities: A review. *Anesthesiology*, 7(6), 599–605. No abstract available.

Bouroche, G., & Bourgain, J. L. (2015). Preoxygenation and general anesthesia: A review. *Minerva Anestesiologica, 81*(8), 910–920. Epub 2015 Jun 5.

Bovill, J. G. (2008). Inhalation anaesthesia: From diethyl ether to xenon. *Handbook of Experimental Pharmacology, 182*, 121–142. https://doi.org/10.1007/978-3-540-74806-9_6. Review.

Brohan, J., & Goudra, B. G. (2017). The role of GABA receptor agonists in anesthesia and sedation. *CNS Drugs, 31*(10), 845–856. https://doi.org/10.1007/s40263-017-0463-7. Review.

Budniewski, A. (1975). Changes in the dynamic compliance coefficient of the lungs and breathing mechanics during administration of diethyl ether and halothane for general anaesthesia with spontaneous respiration. *Anaesthesia, Resuscitation, and Intensive Therapy, 3*(1), 35–41.

Buxbaum, D. M., Yarbrough, G. G., & Carter, M. E. (1973). Biogenic amines and narcotic effects. I. Modification of morphine-induced analgesia and motor activity after alteration of cerebral amine levels. *The Journal of Pharmacology and Experimental Therapeutics, 185*(2), 317–327. No abstract available.

Cascella, M., Bimonte, S., & Muzio, M. R. (2018). Towards a better understanding of anesthesia emergence mechanisms: Research and clinical implications. *World Journal of Methodology, 8*(2), 9–16. https://doi.org/10.5662/wjm.v8.i2.9. eCollection2018 Oct 12. Review.

Chaouloff, F. (1993). Physiopharmacological interactions between stress hormones and central serotonergic systems. *Brain Research Brain Research Reviews, 18*(1), 1–32. Review.

Clausen, N. G., Kähler, S., & Hansen, T. G. (2018). Systematic review of the neurocognitive outcomes used in studies of paediatric anaesthesia neurotoxicity. *British Journal of Anaesthesia, 120*(6), 1255–1273. https://doi.org/10.1016/j.bja.2017.11.107. Epub 2018 Feb 3.

Daniel, S. R., Fry, H. R., & Savord, E. G. (1984). Intravenous "conscious" sedation in periodontal surgery. A selective review and report of 1,708 cases. *The Journal of the Western Society of Periodontology/Periodontal abstracts, 32*(4), 133–146. No abstract available.

De Hert, S. G., Preckel, B., & Schlack, W. S. (2009). Update on inhalational anaesthetics. *Current Opinion in Anaesthesiology, 22*(4), 491–495. https://doi.org/10.1097/ACO.0b013e32832bca38. Review.

Deacon, R. M., & Rawlins, J. N. (1996). Equithesin without chloral hydrate as an anaesthetic for rats. *Psychopharmacology (Berlin), 124*(3), 288–290. Anesthesia changes frequency tuning of neurons in the rat primary auditory cortex. Gaese BH et al. J Neurophysiol. (2001).

Defalque, R. J., & Wright, A. J. (2008). The introduction of ether in Paris revisited. *Bulletin of Anesthesia History, 26*(1), 9–12.

Dey, P. K., & Sharma, H. S. (1983). Ambient temperature and development of traumatic brain oedema in anaesthetized animals. *The Indian Journal of Medical Research*, *77*, 554–563. No abstract available.

Dey, P. K., & Sharma, H. S. (1984). Influence of ambient temperature and drug treatments on brain oedema induced by impact injury on skull in rats. *Indian Journal of Physiology and Pharmacology*, *28*(3), 177–186.

Disma, N., O'Leary, J. D., Loepke, A. W., Brambrink, A. M., Becke, K., Clausen, N. G., et al. (2018). Anesthesia and the developing brain: A way forward for laboratory and clinical research. *Paediatric Anaesthesia*, *28*(9), 758–763. https://doi.org/10.1111/pan.13455. Epub 2018 Aug 16. Review.

Dubey, V., Nau, E., & Sycip, M. (2019). Altered mental status in children after traumatic brain injury. *Pediatric Annals*, *48*(5), e192–e196. https://doi.org/10.3928/19382359-20190422-03. Review.

Dundee, J. W. (1967). Clinical pharmacology of general anesthetics. *Clinical Pharmacology and Therapeutics*, *8*(1), 91–123. Review. No abstract available.

Eger, E. I., 2nd, & Sonner, J. M. (2006). Anaesthesia defined (gentlemen, this is no humbug). *Best Practice & Research. Clinical Anaesthesiology*, *20*(1), 23–29. Review.

England, A., & Jones, R. M. (1992). Inhaled anaesthetic agents: From ether to halothane. *British Journal of Hospital Medicine*, *47*(9), 699–702. Review.

Etsten, B. E., & Shimosato, S. (1966). Influence of stress upon the performance of the heart during halothane and ether anesthesia. *Acta Anaesthesiologica Scandinavica Supplementum*, *23*, 242–247. No abstract available.

Ferreira, A. L., Nunes, C., Mendes, J. G., & Amorim, P. (2019). Do we have today a reliable method to detect the moment of loss of consciousness during induction of general anaesthesia? *Revista Española de Anestesiología y Reanimación*, *66*(2), 93–103. https://doi.org/10.1016/j.redar.2018.06.008. Epub 2018 Aug 1. Review.

Field, K. J., & Lang, C. M. (1988). Hazards of urethane (ethyl carbamate): A review of the literature. *Laboratory Animals*, *22*(3), 255–262. Review.

Finer, B. (1965). Divinyl ether. *British Journal of Anaesthesia*, *37*(9), 661–666. Review. No abstract available.

Franks, N. P., & Honoré, E. (2004). The TREK K2P channels and their role in general anaesthesia and neuroprotection. *Trends in Pharmacological Sciences*, *25*(11), 601–608. Review.

Fuller, R. W. (1981). Serotonergic stimulation of pituitary-adrenocortical function in rats. *Neuroendocrinology*, *32*(2), 118–127. Review. No abstract available.

Gaese, B. H., & Ostwald, J. (2001). Anesthesia changes frequency tuning of neurons in the rat primary auditory cortex. *Journal of Neurophysiology*, *86*(2), 1062–1066.

Gardner, A. J., & Menon, D. K. (2018). Moving to human trials for argon neuroprotection in neurological injury: A narrative review. *British Journal of Anaesthesia*, *120*(3), 453–468. https://doi.org/10.1016/j.bja.2017.10.017. Epub 2017 Nov 24. Review.

Ghosh, M. N., & Banerjee, U. (1965). Effect of premedication on the safety index of ether, chloroform and halothane in mice. I. Atropine, 5-hydroxytryptamine, reserpine and hydergine. *Archives Internationales de Pharmacodynamie et de Thérapie*, *158*(1), 91–96. No abstract available.

Goyagi, T. (2019). Erythropoietin reduces neurodegeneration and long-term memory deficits following sevoflurane exposure in neonatal rats. *Neurotoxicity Research*, 1–10. https://doi.org/10.1007/s12640-019-00028-8. [Epub ahead of print].

Greuter, L., Weil, A. G., Hildebrandt, G., & Surbeck, W. (2017). Intrarectal anesthesia in neurosurgery. *World Neurosurgery*, *104*, 158–160. https://doi.org/10.1016/j.wneu.2017.05.015. Epub 2017 May 11. Review.

Halsey, M. J. (1992). Molecular interactions of anaesthetics with biological membranes. *General Pharmacology*, *23*(6), 1013–1016. Review.

Helmick, K. M., Spells, C. A., Malik, S. Z., Davies, C. A., Marion, D. W., & Hinds, S. R. (2015). Traumatic brain injury in the US military: Epidemiology and key clinical and research programs. *Brain Imaging and Behavior*, *9*(3), 358–366. https://doi.org/10.1007/s11682-015-9399-z.

Hen, J. T., Wei, L., Chen, T. L., Huang, C. J., & Chen, R. M. (2018). Regulation of cytochrome P450 gene expression by ketamine: A review. *Expert Opinion on Drug Metabolism & Toxicology*, *14*(7), 709–720. https://doi.org/10.1080/17425255.2018.1487397. Epub 2018 Jun 17.

Hicks, A., James, A., Spitz, G., & Ponsford, J. (2019). Traumatic brain injury as a risk factor for dementia and Alzheimer's disease: Critical review of study methodologies. *Journal of Neurotrauma*. https://doi.org/10.1089/neu.2018.6346. [Epub ahead of print].

Hingson, R. A. (1961). Review of the uses of general anesthesia. *Transactions of the New England Obstetrical and Gynecological Society*, *15*, 67–78. No abstract available.

Hoetzel, A., & Schmidt, R. (2010). Regulatory role of anesthetics on heme oxygenase-1. *Current Drug Targets*, *11*(12), 1495–1503. Review.

Hwang, H. W., Hyun, H. K., Kim, Y. J., Kim, J. W., & Shin, T. J. (2017). Evaluation of stress response during mesiodens extraction under general anesthesia using heart rate variability. *Journal of Oral and Maxillofacial Surgery*, *75*(5), 931–937. https://doi.org/10.1016/j.joms.2016.10.022. Epub 2016 Oct 28.

Ivankovich, A. D., Miletich, D. J., Grossman, R. K., Albrecht, R. F., El-Etr, A. A., & Cairoli, V. J. (1976). The effect of enflurane, isoflurane, fluroxene, methoxyflurane and diethyl ether anesthesia on ouabain tolerance in the dog. *Anesthesia and Analgesia*, *55*(3), 360–365.

Jevtovic-Todorovic, V., Hartman, R. E., Izumi, Y., Benshoff, N. D., Dikranian, K., Zorumski, C. F., et al. (2003). Early exposure to common anesthetic agents causes widespread neurodegeneration in the developing rat brain and persistent learning deficits. *The Journal of Neuroscience*, *23*(3), 876–882.

Jiang, J., & Jiang, H. (2015). Effect of the inhaled anesthetics isoflurane, sevoflurane and desflurane on the neuropathogenesis of Alzheimer's disease (review). *Molecular Medicine Reports*, *12*(1), 3–12. https://doi.org/10.3892/mmr.2015.3424. Epub 2015 Mar 4. Review.

Jin, Y., Dougherty, S. E., Wood, K., Sun, L., Cudmore, R. H., Abdalla, A., et al. (2016). Regrowth of serotonin axons in the adult mouse brain following injury. *Neuron*, *91*(4), 748–762. https://doi.org/10.1016/j.neuron.2016.07.024. Epub 2016 Aug 4.

Klamt, J. G., Vicente, W. V. A., Garcia, L. V., Carmona, F., Abrão, J., Menardi, A. C., et al. (2017). Neuroprotective anesthesia regimen and intensive management for pediatric cardiac surgery with cardiopulmonary bypass: A review and initial experience. *Brazilian Journal of Cardiovascular Surgery*, *32*(6), 523–529. https://doi.org/10.21470/1678-9741-2016-0064. Review.

Komatsu, H., Nogaya, J., & Ogli, K. (1994). Volatile anaesthetics as central nervous system excitants. *Annals of the Academy of Medicine, Singapore*, *23*(6 Suppl), 130–138. Review.

Kopp Lugli, A., Yost, C. S., & Kindler, C. H. (2009). Anaesthetic mechanisms: Update on the challenge of unravelling the mystery of anaesthesia. *European Journal of Anaesthesiology*, *26*(10), 807–820. https://doi.org/10.1097/EJA.0b013e32832d6b0f. Review.

Korneyev, A. Y., Costa, E., & Guidotti, A. (1993). During anesthetic-induced activation of hypothalamic pituitary adrenal axis, blood-borne steroids fail to contribute to the anesthetic effect. *Neuroendocrinology*, *57*(3), 559–565.

Kosari-Nasab, M., Shokouhi, G., Azarfarin, M., Bannazadeh Amirkhiz, M., Mesgari Abbasi, M., & Salari, A. A. (2019). Serotonin 5-HT1A receptors modulate depression-related symptoms following mild traumatic brain injury in male adult mice. *Metabolic Brain Disease*, *34*(2), 575–582. https://doi.org/10.1007/s11011-018-0366-4. Epub 2019 Jan 3.

Kunst, G., & Klein, A. A. (2015). Peri-operative anaesthetic myocardial preconditioning and protection—Cellular mechanisms and clinical relevance in cardiac anaesthesia. *Anaesthesia*, *70*(4), 467–482. https://doi.org/10.1111/anae.12975.

Laaksonen, L., Kallioinen, M., Långsjö, J., Laitio, T., Scheinin, A., Scheinin, J., et al. (2018). Comparative effects of dexmedetomidine, propofol, sevoflurane, and S-ketamine on regional cerebral glucose metabolism in humans: A positron emission tomography study. *British Journal of Anaesthesia*, *121*(1), 281–290. https://doi.org/10.1016/j.bja.2018.04.008. Epub 2018 May 8.

Lafuente, J. V., Sharma, A., Muresanu, D. F., Ozkizilcik, A., Tian, Z. R., Patnaik, R., et al. (2018). Repeated forced swim exacerbates methamphetamine-induced neurotoxicity: Neuroprotective effects of nanowired delivery of 5-HT3-receptor antagonist ondansetron. *Molecular Neurobiology*, *55*(1), 322–334. https://doi.org/10.1007/s12035-017-0744-7.

Learned, L. O. (1949). A general review of nitrous oxide anesthesia. *Rocky Mountain Medical Journal*, *46*(3), 201–205. No abstract available.

Li, Y., Dong, H., Tan, S., Qian, Y., & Jin, W. (2019). Effects of thoracic epidural anesthesia/analgesia on the stress response, pain relief, hospital stay, and treatment costs of patients with esophageal carcinoma undergoing thoracic surgery: A single-center, randomized controlled trial. *Medicine (Baltimore)*, *98*(7). e14362. https://doi.org/10.1097/MD.0000000000014362.

Li, Y., Li, X., Zhao, J., Li, L., Wang, Y., Zhang, Y., et al. (2018). Midazolam attenuates autophagy and apoptosis caused by ketamine by decreasing reactive oxygen species in the hippocampus of fetal rats. *Neuroscience*, *388*, 460–471. https://doi.org/10.1016/j.neuroscience.2018.03.040. Epub 2018 Apr 7.

Li, Q., Wang, P., Huang, C., Chen, B., Liu, J., Zhao, M., et al. (2019). N-acetyl serotonin protects neural progenitor cells against oxidative stress-induced apoptosis and improves neurogenesis in adult mouse hippocampus following traumatic brain injury. *Journal of Molecular Neuroscience*, *67*(4), 574–588. https://doi.org/10.1007/s12031-019-01263-6. Epub 2019 Jan 25.

Liepert, J. (2016). Update on pharmacotherapy for stroke and traumatic brain injury recovery during rehabilitation. *Current Opinion in Neurology*, *29*(6), 700–705.

Lowenstein, D. H., Aminoff, M. J., & Simon, R. P. (1988). Barbiturate anesthesia in the treatment of status epilepticus: Clinical experience with 14 patients. *Neurology*, *38*(3), 395–400.

Maggi, C. A., & Meli, A. (1986a). Suitability of urethane anesthesia for physiopharmacological investigations. Part 3: Other systems and conclusions. *Experientia*, *42*(5), 531–537. Review.

Maggi, C. A., & Meli, A. (1986b). Suitability of urethane anesthesia for physiopharmacological investigations in various systems. Part 2: Cardiovascular system. *Experientia*, *42*(3), 292–297. Review.

Maggi, C. A., & Meli, A. (1986c). Suitability of urethane anesthesia for physiopharmacological investigations in various systems. Part 1: General considerations. *Experientia*, *42*(2), 109–114. Review.

Malviya, S., Voepel-Lewis, T., Tait, A. R., Reynolds, P. I., Gujar, S. K., Gebarski, S. S., et al. (2004). Pentobarbital vs chloral hydrate for sedation of children undergoing MRI: Efficacy and recovery characteristics. *Paediatric Anaesthesia*, *14*(7), 589–595.

Manell, E., Jensen-Waern, M., & Hedenqvist, P. (2017). Anaesthesia and changes in parameters that reflect glucose metabolism in pigs—A pilot study. *Laboratory Animals*, *51*(5), 509–517. https://doi.org/10.1177/0023677216682773. Epub 2016 Dec 15.

Mantz, J., & Dahmani, S. (2008). Exploring the mechanisms of general anesthesia. *Acta Anaesthesiologica Belgica*, *59*(3), 167–171. Review. No abstract available.

Marchesini, V., & Disma, N. (2019). Anaesthetic neuroprotection in children: Does it exist or is it all just bad? *Current Opinion in Anaesthesiology, 32*(3), 363–369. https://doi.org/10.1097/ACO.0000000000000723.

Mohanty, S., Dey, P. K., & Ray, A. K. (1979). The role of serotonin in cerebral oedema. *The Indian Journal of Medical Research, 69*, 1001–1007. No abstract available.

Mohanty, S., Dey, P. K., Sen, P. C., & Ray, A. (1978). Accumulation of serotonin in human cerebral contusion. *Neurology India, 26*(2), 68–70. No abstract available.

Mohanty, S., & Mazumdar, S. (1978). Role of serotonin in human cerebral oedema and contusion. *The Indian Journal of Medical Research, 67*, 1029–1032. No abstract available.

Moore, R. Y., Wong, S. L., & Heller, A. (1965). Regional effects of hypothalamic lesions on brain serotonin. *Archives of Neurology, 13*(4), 346–354. No abstract available.

Mueller, R. A., & Hunt, R. (1998). Antagonism of ketamine-induced anesthesia by an inhibitor of nitric oxide synthesis: A pharmacokinetic explanation. *Pharmacology, Biochemistry, and Behavior, 60*(1), 15–22.

Muir, W. W. (2010). NMDA receptor antagonists and pain: Ketamine. *The Veterinary Clinics of North America Equine Practice, 26*(3), 565–578. https://doi.org/10.1016/j.cveq.2010.07.009. Review.

Murnane, K. S. (2019). Serotonin 2A receptors are a stress response system: Implications for post-traumatic stress disorder. *Behavioural Pharmacology, 30*(2 and 3—Special Issue), 151–162. https://doi.org/10.1097/FBP.0000000000000459.

Musizza, B., & Ribaric, S. (2010). Monitoring the depth of anaesthesia. *Sensors (Basel), 10*(12), 10896–10935. https://doi.org/10.3390/s101210896. Epub 2010 Dec 3. Review.

Narahashi, T. (1991). Transmitter-activated ion channels as the target of chemical agents. *Advances in Experimental Medicine and Biology, 287*, 61–73. Review.

Natarajan, R., Northrop, N. A., & Yamamoto, B. K. (2015). Protracted effects of chronic stress on serotonin-dependent thermoregulation. *Stress, 18*(6), 668–676. https://doi.org/10.3109/10253890.2015.1087502. Epub 2015 Sep 28.

Nicoletti, A., Vasta, R., Mostile, G., Nicoletti, G., Arabia, G., Iliceto, G., et al. (2017). Head trauma and Parkinson's disease: Results from an Italian case-control study. *Neurological Sciences, 38*(10), 1835–1839. https://doi.org/10.1007/s10072-017-3076-5. Epub 2017 Jul 26.

No Author Listed. (1852). Anæsthesia, or the employment of chloroform and ether in surgery, midwifery, &c. *The British and Foreign Medico-Chirurgical Review, 9*(18), 443–467. Review. No abstract available.

O'Brien, J. J. (1967). Recent advances in general anaesthesia. *Australian Veterinary Journal, 43*(10), 472–477. Review. No abstract available.

Okamoto, M. (1984). Barbiturate tolerance and physical dependence: Contribution of pharmacological factors. *NIDA Research Monograph, 54*, 333–347. Review. No abstract available.

Oliveira, C. R., Bernardo, W. M., & Nunes, V. M. (2017). Benefit of general anesthesia monitored by bispectral index compared with monitoring guided only by clinical parameters. Systematic review and meta-analysis. *Brazilian Journal of Anesthesiology, 67*(1), 72–84. https://doi.org/10.1016/j.bjane.2015.09.001. Epub 2016 Apr 14.

Olsen, R. W., & Li, G. D. (2011). GABA(A) receptors as molecular targets of general anesthetics: Identification of binding sites provides clues to allosteric modulation. *Canadian Journal of Anaesthesia, 58*(2), 206–215. https://doi.org/10.1007/s12630-010-9429-7. Epub 2010 Dec 31. Review.

Overgaard, A., Lieblich, S. E., Richardson, R., Galea, L. A. M., & Frokjaer, V. G. (2018). Paroxetine blunts the corticosterone response to swim-induced stress and increases depressive-like behavior in a rat model of postpartum depression. *Psychoneuroendocrinology, 89*, 223–228. https://doi.org/10.1016/j.psyneuen.2017.10.021. Epub 2017 Oct 31.

Pagliardini, S., Funk, G. D., & Dickson, C. T. (2013). Breathing and brain state: Urethane anesthesia as a model for natural sleep. *Respiratory Physiology & Neurobiology, 188*(3), 324–332. https://doi.org/10.1016/j.resp.2013.05.035. Epub 2013 Jun 7. Review.

Park, T. Y., Nishida, K. S., Wilson, C. M., Jaiswal, S., Scott, J., Hoy, A. R., et al. (2017). Effects of isoflurane anesthesia and intravenous morphine self-administration on regional glucose metabolism ([^{18} F]FDG-PET) of male Sprague-Dawley rats. *The European Journal of Neuroscience, 45*(7), 922–931. https://doi.org/10.1111/ejn.13542. Epub 2017 Mar 4.

Pasternak, J. J., & Lanier, W. L. (2007). Neuroanesthesiology review–2006. *Journal of Neurosurgical Anesthesiology, 19*(2), 70–92. Review.

Pattinson, C. L., Shahim, P., Taylor, P., Dunbar, K., Guedes, V. A., Motamedi, V., et al. (2019). Elevated tau in military personnel relates to chronic symptoms following traumatic brain injury. *The Journal of Head Trauma Rehabilitation*. https://doi.org/10.1097/HTR.0000000000000485. [Epub ahead of print].

Pekny, T., Andersson, D., Wilhelmsson, U., Pekna, M., & Pekny, M. (2014). Short general anaesthesia induces prolonged changes in gene expression in the mouse hippocampus. *Acta Anaesthesiologica Scandinavica, 58*(9), 1127–1133. https://doi.org/10.1111/aas.12369. Epub 2014 Jul 17.

Perouansky, M. (2008). General anesthetics and long-term neurotoxicity. *Handbook of Experimental Pharmacology, 182*, 143–157. https://doi.org/10.1007/978-3-540-74806-9_7. Review.

Petrenko, A. B., Yamakura, T., Sakimura, K., & Baba, H. (2014). Defining the role of NMDA receptors in anesthesia: Are we there yet? *European Journal of Pharmacology, 723*, 29–37. https://doi.org/10.1016/j.ejphar.2013.11.039. Epub 2013 Dec 11. Review.

Pettersson, L. O., & Akerman, B. (1979). Effect of different types of anaesthesia including percutaneous local anaesthesia on survival of experimental skin flaps. *Scandinavian Journal of Plastic and Reconstructive Surgery, 13*(2), 237–240.

Pieper, J., Chang, D. G., Mahasin, S. Z., Swan, A. R., Quinto, A. A., Nichols, S. L., et al. (2019). Brain amygdala volume increases in veterans and active-duty military personnel with combat-related posttraumatic stress disorder and mild traumatic brain injury. *The Journal of Head Trauma Rehabilitation*. https://doi.org/10.1097/HTR.0000000000000492. [Epub ahead of print].

Pocock, G., & Richards, C. D. (1993). Excitatory and inhibitory synaptic mechanisms in anaesthesia. *British Journal of Anaesthesia, 71*(1), 134–147. Review.

Ptyushkin, P., Vidmar, G., Burger, H., & Marincek, C. (2012). Use of the international classification of functioning, disability, and health in traumatic brain injury rehabilitation: Linking issues and general perspectives. *American Journal of Physical Medicine & Rehabilitation, 91*(13 Suppl 1), S48–S54. https://doi.org/10.1097/PHM.0b013e31823d4e99.

Puglisi-Allegra, S., & Andolina, D. (2015). Serotonin and stress coping. *Behavioural Brain Research, 277*, 58–67. https://doi.org/10.1016/j.bbr.2014.07.052. Epub 2014 Aug 6.

Riegelhaupt, P. M., Tibbs, G. R., & Goldstein, P. A. (2018). HCN and K_{2P} channels in anesthetic mechanisms research. *Methods in Enzymology, 602*, 391–416. https://doi.org/10.1016/bs.mie.2018.01.015. Epub 2018 Mar 2. Review.

RM, Zilberstein (1960). Effects of reserpine, serotonin and vasopressin on the survival of cold-stressed rats. *Nature, 185*, 249. No abstract available.

Roberts, M., & Jagdish, S. A. (2016). History of intravenous anesthesia in war (1656–1988). *Journal of Anesthesia History, 2*(1), 13–21. https://doi.org/10.1016/j.janh.2015.10.007. Epub 2015 Nov 11.

Robinson, D. H., & Toledo, A. H. (2012). Historical development of modern anesthesia. *Journal of Investigative Surgery, 25*(3), 141–149. https://doi.org/10.3109/08941939.2012.690328.

Rona, R. J., Jones, M., Jones, N., Fear, N. T., & Wessely, S. (2019). Long-term correlates of mild traumatic brain injury on postconcussion symptoms after deployment to Iraq and Afghanistan in the UK military. *The Journal of Head Trauma Rehabilitation*, 1–11. https://doi.org/10.1097/HTR.0000000000000497. [Epub ahead of print].

Rudelli, R., Strom, J. O., Welch, P. T., & Ambler, M. W. (1982). Posttraumatic premature Alzheimer's disease. Neuropathologic findings and pathogenetic considerations. *Archives of Neurology*, *39*(9), 570–575.

Saari, T. I., Uusi-Oukari, M., Ahonen, J., & Olkkola, K. T. (2011). Enhancement of GABAergic activity: Neuropharmacological effects of benzodiazepines and therapeutic use in anesthesiology. *Pharmacological Reviews*, *63*(1), 243–267. https://doi.org/10.1124/pr.110.002717. Epub 2011 Jan 18. Review.

Sabourin, V. M., Holland, R., Mau, C., Gandhi, C. D., & Prestigiacomo, C. J. (2016). Head injury in heroes of the civil war and its lasting influence. *Neurosurgical Focus*, *41*(1), E4. https://doi.org/10.3171/2016.3.FOCUS1586.

Sano, Y., Ito, S., Yoneda, M., Nagasawa, K., Matsuura, N., Yamada, Y., et al. (2016). Effects of various types of anesthesia on hemodynamics, cardiac function, and glucose and lipid metabolism in rats. *American Journal of Physiology. Heart and Circulatory Physiology*, *311*(6), H1360–H1366. https://doi.org/10.1152/ajpheart.00181.2016. Epub 2016 Sep 30.

Sepúlveda, P. O., Tapia, L. F., & Monsalves, S. (2019). Neural inertia and differences between loss of and recovery from consciousness during total intravenous anaesthesia: A narrative review. *Anaesthesia*, *74*(6), 801–809. https://doi.org/10.1111/anae.14609. Epub 2019 Mar 5. Review.

Sharma, H. S. (1987). Effect of captopril (a converting enzyme inhibitor) on blood-brain barrier permeability and cerebral blood flow in normotensive rats. *Neuropharmacology*, *26*(1), 85–92.

Sharma, H. S. (2004a). Influence of serotonin on the blood-brain and blood-spinal cord barriers. In H. S. Sharma & J. Westman (Eds.), *The blood-spinal cord and brain barriers in health and disease* (pp. 117–158). San Diego: Elsevier Academic Press.

Sharma, H. S. (2004b). Blood-brain and spinal cord barriers in stress. In H. S. Sharma & J. Westman (Eds.), *The blood-spinal cord and brain barriers in health and disease* (pp. 231–298). San Diego: Elsevier Academic Press.

Sharma, H. S. (2006). Hyperthermia influences excitatory and inhibitory amino acid neurotransmitters in the central nervous system. An experimental study in the rat using behavioural, biochemical, pharmacological, and morphological approaches. *Journal of Neural Transmission (Vienna)*, *113*(4), 497–519.

Sharma, H. S. (2007). Methods to produce hyperthermia-induced brain dysfunction. *Progress in Brain Research*, *162*, 173–199. Review.

Sharma, H. S., Ali, S. F., Tian, Z. R., Hussain, S. M., Schlager, J. J., Sjöquist, P. O., et al. (2009). Chronic treatment with nanoparticles exacerbate hyperthermia induced blood-brain barrier breakdown, cognitive dysfunction and brain pathology in the rat. Neuroprotective effects of nanowired-antioxidant compound H-290/51. *Journal of Nanoscience and Nanotechnology*, *9*(8), 5073–5090.

Sharma, H. S., & Cervós-Navarro, J. (1990). Brain oedema and cellular changes induced by acute heat stress in young rats. *Acta Neurochirurgica. Supplementum*, *51*, 383–386.

Sharma, H. S., Cervós-Navarro, J., & Dey, P. K. (1991a). Acute heat exposure causes cellular alteration in cerebral cortex of young rats. *Neuroreport*, *2*(3), 155–158.

Sharma, H. S., Cervós-Navarro, J., & Dey, P. K. (1991b). Increased blood-brain barrier permeability following acute short-term swimming exercise in conscious normotensive young rats. *Neuroscience Research*, *10*(3), 211–221.

Sharma, H. S., & Dey, P. K. (1981). Impairment of blood-brain barrier (BBB) in rat by immobilization stress: Role of serotonin (5-HT). *Indian Journal of Physiology and Pharmacology*, *25*(2), 111–122.

Sharma, H. S., & Dey, P. K. (1986a). Probable involvement of 5-hydroxytryptamine in increased permeability of blood-brain barrier under heat stress in young rats. *Neuropharmacology*, *25*(2), 161–167.

Sharma, H. S., & Dey, P. K. (1986b). Influence of long-term immobilization stress on regional blood-brain barrier permeability, cerebral blood flow and 5-HT level in conscious normotensive young rats. *Journal of the Neurological Sciences*, *72*(1), 61–76.

Sharma, H. S., & Dey, P. K. (1988). EEG changes following increased blood-brain barrier permeability under long-term immobilization stress in young rats. *Neuroscience Research*, *5*(3), 224–239.

Sharma, H. S., Dey, P. K., & Olsson, Y. (1989). Brain edema, blood-brain barrier permeability and cerebral blood flow changes following intracarotid infusion of serotonin: modification with cyproheptadine and indomethacin. In *Pharmacology of cerebral Ischemia 1988* (pp. 317–323). Ed J Krieglstein: CRC Press, Boca Raton, Florida.

Sharma, H. S., Olsson, Y., & Cervós-Navarro, J. (1993). p-Chlorophenylalanine, a serotonin synthesis inhibitor, reduces the response of glial fibrillary acidic protein induced by trauma to the spinal cord. An immunohistochemical investigation in the rat. *Acta Neuropathologica*, *86*(5), 422–427.

Sharma, H. S., Olsson, Y., & Dey, P. K. (1989). Influence of serotonin on regional blood-brain barrier permeability and cerebral blood flow in the rat. In J. Seylaz & E. T. MacKenzie (Eds.), *Neurotransmission and cerebrovascular function I* (pp. 253–255). Amsterdam: Elsevier.

Sharma, H. S., Olsson, Y., & Dey, P. K. (1990). Changes in blood-brain barrier and cerebral blood flow following elevation of circulating serotonin level in anesthetized rats. *Brain Research*, *517*(1-2), 215–223.

Sharma, H. S., Patnaik, R., Patnaik, S., Mohanty, S., Sharma, A., & Vannemreddy, P. (2007). Antibodies to serotonin attenuate closed head injury induced blood brain barrier disruption and brain pathology. *Annals of the New York Academy of Sciences*, *1122*, 295–312.

Sharma, H. S., Patnaik, R., Patnaik, S., Sharma, A., Mohanty, S., & Vannemreddy, P. (2010). Antibodies to dynorphin a (1-17) attenuate closed head injury induced blood-brain barrier disruption, brain edema formation and brain pathology in the rat. *Acta Neurochirurgica. Supplement*, *106*, 301–306. https://doi.org/10.1007/978-3-211-98811-4_56.

Sharma, H. S., & Sharma, A. (2007). Nanoparticles aggravate heat stress induced cognitive deficits, blood-brain barrier disruption, edema formation and brain pathology. *Progress in Brain Research*, *162*, 245–273. Review.

Sharma, H. S., & Sjöquist, P. O. (2002). A new antioxidant compound H-290/51 modulates glutamate and GABA immunoreactivity in the rat spinal cord following trauma. *Amino Acids*, *23*(1-3), 261–272.

Sharma, H. S., Westman, J., Navarro, J. C., Dey, P. K., & Nyberg, F. (1995). Probable involvement of serotonin in the increased permeability of the blood-brain barrier by forced swimming. An experimental study using Evans blue and 131I-sodium tracers in the rat. *Behavioural Brain Research*, *72*(1-2), 189–196.

Sharma, H. S., Westman, J., & Nyberg, F. (1998). Pathophysiology of brain edema and cell changes following hyperthermic brain injury. *Progress in Brain Research*, *115*, 351–412. Review. No abstract available.

Sharma, H. S., Winkler, T., Stålberg, E., Mohanty, S., & Westman, J. (2000). p-Chlorophenylalanine, an inhibitor of serotonin synthesis reduces blood-brain barrier permeability, cerebral blood flow, edema formation and cell injury following trauma to the rat brain. *Acta Neurochirurgica. Supplement*, *76*, 91–95.

Sharma, H. S., Zimmer, C., Westman, J., & Cervós-Navarro, J. (1992). Acute systemic heat stress increases glial fibrillary acidic protein immunoreactivity in brain: Experimental observations in conscious normotensive young rats. *Neuroscience*, *48*(4), 889–901.

Slupe, A. M., & Kirsch, J. R. (2018). Effects of anesthesia on cerebral blood flow, metabolism, and neuroprotection. *Journal of Cerebral Blood Flow and Metabolism, 38*(12), 2192–2208. https://doi.org/10.1177/0271678X18789273. Epub 2018 Jul 16.

Smith, L. M., Cozowicz, C., Uda, Y., Memtsoudis, S. G., & Barrington, M. J. (2017). Neuraxial and combined neuraxial/general anesthesia compared to general anesthesia for major truncal and lower limb surgery: A systematic review and meta-analysis. *Anesthesia and Analgesia, 125*(6), 1931–1945. https://doi.org/10.1213/ANE.0000000000002069.

Son, Y. (2010). Molecular mechanisms of general anesthesia. *Korean Journal of Anesthesiology, 59*(1), 3–8. https://doi.org/10.4097/kjae.2010.59.1.3. Epub 2010 Jul 21.

Soplata, A. E., McCarthy, M. M., Sherfey, J., Lee, S., Purdon, P. L., Brown, E. N., et al. (2017). Thalamocortical control of propofol phase-amplitude coupling. *PLoS Computational Biology, 13*(12), e1005879. https://doi.org/10.1371/journal.pcbi.1005879. eCollection 2017 Dec.

Stein, D. G. (2019). How much can we ask of molecular biomarkers? An opinion on Pattinson et al, "elevated tau in military personnel relates to chronic symptoms following traumatic brain injury" *The Journal of Head Trauma Rehabilitation*. https://doi.org/10.1097/HTR.0000000000000494. [Epub ahead of print]. No abstract available.

Steinberg, E. A., Wafford, K. A., Brickley, S. G., Franks, N. P., & Wisden, W. (2015). The role of K_2p channels in anaesthesia and sleep. *Pflügers Archiv, 467*(5), 907–916. https://doi.org/10.1007/s00424-014-1654-4. Epub 2014 Dec 9. Review.

Takizawa, C., Gemmell, E., Kenworthy, J., & Speyer, R. A. (2016). Systematic review of the prevalence of oropharyngeal dysphagia in stroke, Parkinson's disease, Alzheimer's disease, head injury, and pneumonia. *Dysphagia, 31*(3), 434–441. https://doi.org/10.1007/s00455-016-9695-9. Epub 2016 Mar 12.

Trapani, G., Altomare, C., Liso, G., Sanna, E., & Biggio, G. (2000). Propofol in anesthesia. Mechanism of action, structure-activity relationships, and drug delivery. *Current Medicinal Chemistry*, 7(2), 249–271. Review.

Troup, G. R. (1947). A review of curare as an adjuvant to general anaesthesia. *The Medical Journal of Australia, 2*(9), 259–261. No abstract available.

Van Biesen, S., Van de Velde, M., & Rex, S. (2015). Anesthesia and neurotoxicity in the developing brain: A non-systematic review. *Acta Anaesthesiologica Belgica, 66*(3), 67–79. Review.

Vannemreddy, P., Ray, A. K., Patnaik, R., Patnaik, S., Mohanty, S., & Sharma, H. S. (2006). Zinc protoporphyrin IX attenuates closed head injury-induced edema formation, blood-brain barrier disruption, and serotonin levels in the rat. *Acta Neurochirurgica Supplement, 96*, 151–156.

Villars, P. S., Kanusky, J. T., & Dougherty, T. B. (2004). Stunning the neural nexus: Mechanisms of general anesthesia. *AANA Journal, 72*(3), 197–205. Review.

Vutskits, L., & Davidson, A. (2017). Update on developmental anesthesia neurotoxicity. *Current Opinion in Anaesthesiology, 30*(3), 337–342. https://doi.org/10.1097/ACO.0000000000000461. Review.

Wagner, R. I., Frid, I. A., Evtukhin, A. I., Barchuk, A. S., & Shulga, N. I. (1981). On the end results of surgical treatment of lung cancer patients depending on the type of anesthetic agent. *Journal of Surgical Oncology, 18*(1), 39–46.

Wahl, M., Unterberg, A., Baethmann, A., & Schilling, L. (1988). Mediators of blood-brain barrier dysfunction and formation of vasogenic brain edema. *Journal of Cerebral Blood Flow and Metabolism, 8*(5), 621–634. Review. No abstract available.

Wahl, M., Unterberg, A., Baethmann, A., & Schilling, L. (1990). Mediators of vasogenic brain edema. *Journal of Basic and Clinical Physiology and Pharmacology, 1*(1-4), 221–233. Review. No abstract available.

Wali, F. A. (1985). General anaesthetics: Effects on transmitter release. *General Pharmacology, 16*(1), 7–11. Review.

Watrous, W. G., Davis, F. E., & Anderson, B. M. (1950a). Manually assisted and controlled respiration: Its use during inhalation anesthesia for the maintenance of a near-normal physiologic state: A review. *Anesthesiology, 11*(5), 538–561contd. No abstract available.

Watrous, W. G., Davis, F. E., & Anderson, B. M. (1950b). Manually assisted and controlled respiration; its use during inhalation anesthesia for the maintenance of a near-normal physiologic state; a review. Part II. Controlled and assisted respiration. *Anesthesiology, 11*(6), 661–685. No abstract available.

Wink, J., Veering, B. T., Aarts, L. P. H. J., & Wouters, P. F. (2019). Effects of thoracic epidural anesthesia on neuronal cardiac regulation and cardiac function. *Anesthesiology, 130*(3), 472–491. https://doi.org/10.1097/ALN.0000000000002558.

Wood, R. L. (2008). Long-term outcome of serious traumatic brain injury. *European Journal of Anaesthesiology. Supplement, 42*, 115–122. https://doi.org/10.1017/S0265021507003432. Review.

Wu, L., Zhao, H., Weng, H., & Ma, D. (2019). Lasting effects of general anesthetics on the brain in the young and elderly: "Mixed picture" of neurotoxicity, neuroprotection and cognitive impairment. *Journal of Anesthesia, 33*(2), 321–335. https://doi.org/10.1007/s00540-019-02623-7. Epub 2019 Mar 11. Review.

Xia, Z., Li, H., & Irwin, M. G. (2016). Myocardial ischaemia reperfusion injury: The challenge of translating ischaemic and anaesthetic protection from animal models to humans. *British Journal of Anaesthesia, 117*(Suppl 2), ii44–ii62. https://doi.org/10.1093/bja/aew267. Review.

Yan, J., & Jiang, H. (2014). Dual effects of ketamine: Neurotoxicity versus neuroprotection in anesthesia for the developing brain. *Journal of Neurosurgical Anesthesiology, 26*(2), 155–160. https://doi.org/10.1097/ANA.0000000000000027. Review.

Zgleszewski, S. E., Zurakowski, D., Fontaine, P. J., D'Angelo, M., & Mason, K. P. (2008). Is propofol a safe alternative to pentobarbital for sedation during pediatric diagnostic CT? *Radiology, 247*(2), 528–534. https://doi.org/10.1148/radiol.2472062087.

Zhang, S., Hu, X., Guan, W., Luan, L., Li, B., Tang, Q., et al. (2017). Isoflurane anesthesia promotes cognitive impairment by inducing expression of β-amyloid protein-related factors in the hippocampus of aged rats. *PLoS One, 12*(4). e0175654https://doi.org/10.1371/journal.pone.0175654eCollection 2017.

Zhou, R., & Bickler, P. (2017). Interaction of isoflurane, tumor necrosis factor-α and β-amyloid on long-term potentiation in rat hippocampal slices. *Anesthesia and Analgesia, 124*(2), 582–587. https://doi.org/10.1213/ANE.0000000000001698.

CHAPTER THREE

Potentiation of spinal cord conduction and neuroprotection following nanodelivery of DL-3-*n*-butylphthalide in titanium implanted nanomaterial in a focal spinal cord injury induced functional outcome, blood-spinal cord barrier breakdown and edema formation

Seaab Sahib[a], Feng Niu[b], Aruna Sharma[c], Lianyuan Feng[d], Z. Ryan Tian[a], Dafin Fior Muresanu[e,f], Ala Nozari[g], Hari Shanker Sharma[c,*]

[a]Cell & Molecular Biology Program (CEMB), Department of Chemistry and Biochemistry, University of Arkansas—Fayetteville, Fayetteville, AR, United States
[b]CSPC NBP Pharmaceutical Medicine, Shijiazhuang, China
[c]International Experimental Central Nervous System Injury & Repair (IECNSIR), Department of Surgical Sciences, Anesthesiology & Intensive Care Medicine, Uppsala University Hospital, Uppsala University, Uppsala, Sweden
[d]Department of Neurology, Bethune International Peace Hospital, Shijiazhuang, China
[e]Department of Clinical Neurosciences, University of Medicine & Pharmacy, Cluj-Napoca, Romania
[f]"RoNeuro" Institute for Neurological Research and Diagnostic, Cluj-Napoca, Romania
[g]Anesthesiology & Intensive Care, Massachusetts General Hospital, Boston, MA, United States
*Corresponding author: e-mail addresses: sharma@surgsci.uu.se; harishanker_sharma55@icloud.com; hssharma@aol.com

Contents

International Review of Neurobiology, Volume 146
ISSN 0074-7742
https://doi.org/10.1016/bs.irn.2019.06.009

Abstract

Spinal cord injury (SCI) is a devastating disease inflicting lifetime disability to the victims. Military personnel are quite often victims of SCI for which no suitable therapeutic strategies have been developed so far. The main reason for SCI induced disability is loss of neural connections below and above the lesion site causing motor paralysis and somatosensory disturbances Loss of neuronal connections thwart spinal cord conduction resulting in motor function disability. To enhance spinal cord conduction grafting of peripheral nerves, implant of hydrogels filled with neuroprotective drugs is used but so far, no satisfactory results re achieved. In this regards implants of microelectrode for enhancing tissue connectivity is suggested that is still under experimental state. We have used titanium implant with or without TiO_2 nanowires in a focal spinal cord injury and studies spinal cord pathology and motor function. In addition, we also combined with nanowired delivery of a potential neuroprotective drug DL-3-*n*-butylphthalide (DL-NBP) to the spinal cord in a rat model. Our observations show that a combination of titanium implant with nanowired delivery of DL-NBP induces superior neuroprotection and enhance motor functions after SCI. This treatment also restored blood-spinal cord barrier (BSCB) function and reduces edema formation and cell injury after SCI, not reports earlier.

1. Introduction

Spinal cord injury (SCI) induces lifetime disability due to breakdown of the spinal neuronal networks (Anderson et al., 2008; Ditunno Jr., 1997; Gardner, Theocleous, & Krishnan, 1988; Ma, Chan, & Carruthers, 2014; New, 2005; Nyberg, Sharma, & Wisenfeld-Hallin, 1995; Sharma, 2008;

Stålberg, Sharma, & Olsson, 1998; Wood-Dauphinée et al., 2002; Woolsey, 1988). Depending on the magnitude and severity of spinal cord trauma either caused by contusion, incision, transaction leading to either complete or partial SCI (Dietz, Wirz, & Jensen, 1997; Dimitrijevic, Hsu, & McKay, 1992; Kakulas, 1988; Meinecke, 1985; Tator & Fehlings, 1991; Whelan, Halpine, Christie, & McVeigh, 2018). In such situations impulses from brain do not travel below the lesion causing paralysis (Blight, 1988; Dimitrijevic, Danner, & Mayr, 2015; Ghosh & Pearse, 2015; McKay, Stokic, & Dimitrijevic, 1997; Minassian & Hofstoetter, 2016; Sadowsky, 2001; Tidoni, Tieri, & Aglioti, 2015). Also, stimulation of sciatic nerves does not allow electrical impulses to travel beyond the lesion site if the cord injury is complete (Dimitrijevic, Faganel, & Young, 1981; Ievins & Moritz, 2017; Thomaz et al., 2019; Wilson et al., 2019). Several attempts have been done to promote regeneration of the damaged spinal cord in order to revive axonal connection for nerve impulses to travel both above and below the lesion (Hughes, 1984; Krishna et al., 2013; Puchala & Windle, 1977; Schwab, 1992; Tator, 1972; Veneruso et al., 2019; Yu et al., 2019). However, this task is so far still not complete and the spinal cord conduction after injury is very hard to revive (de la Torre, 1981; Gavaret et al., 2017; Grundy & Friedman, 1987; Nash Jr & Brown, 1989; Nuwer, 1988; Schramm, 1985).

To accomplish this, lost connections between axons and neurons several studies examined the possibility to connect the severed ends with hydrogels filled with drug for regeneration of axonal growth beyond the lesion site (Chedly et al., 2017; Führmann et al., 2016, 2017; Hong et al., 2017; Pakulska, Ballios, & Shoichet, 2012). However, in most studies axonal growth is largely limited to few 100 μm and all connections are not established within the cord after injury (Dietz & Schwab, 2017; Schnell & Schwab, 1990, 1993; Schwab, 1993; Theisen et al., 2017). Another way to bridge the gap of lost tissues and connections within the spinal cord after injury is to use peripheral nerve grafts or fetal tissue that are somehow showed to maintain neuronal connections within 7–8 weeks of implantation but full recovery of spinal cord tissues and its ascending and descending connections leading to normal walking is far beyond the reach (Bunge & Wood, 2012; Côté, Amin, Tom, & Houle, 2011; Gad et al., 2013; Nógrádi, Pajer, & Márton, 2011; Yang, He, & Hao, 2015).

Recently, implantations of miniature titanium electrodes that are flexible enough are implanted into the cord tissue that can be electrically stimulated for neuronal excitation enough to walk in laboratory animals (Cheng, Zhang, & Fu, 2019; Hurlbert, Tator, & Theriault, 1993;

Lee et al., 2011). However, still there is a long way to go in order to fully establish electrical connections between ascending and descending neurons and axons within the spinal cord after injury. The electrical impulses within the spinal cord either spontaneous or evoked represent the health of the spinal cord functions (Cheng et al., 2019; de la Torre, 1981; Grundy & Friedman, 1987; Hu, Jeffery, & Granger, 2018; Nash Jr & Brown, 1989; Nuwer, 1988; Winkler, Sharma, Stålberg, & Westman, 1998). Thus, spinal cord evoked potentials (SCEP) are good measures of spinal cord pathology and edema formation after SCI (Deecke & Tator, 1973; Martin & Bloedel, 1973; Mesbah, Angeli, Keynton, El-Baz, & Harkema, 2017; Perot Jr, 1973; Thirumala et al., 2016). Thus, these SCEP and somatosensory evoked potentials (SSEP) are good measure for predicting spinal cord pathology and motor or sensory functions (see Deecke & Tator, 1973; Ducker, Salcman, Lucas, Garrison, & Perot Jr., 1978; Kobrine, Evans, & Rizzoli, 1978; Kojima, Yamamoto, Ogino, Okada, & Ono, 1979; Martin & Bloedel, 1973; Mesbah et al., 2017; Perot Jr, 1973; Rowed, McLean, & Tator, 1978; Stålberg et al., 1998; Tamaki, Tsuji, Inoue, & Kobayashi, 1981; Thirumala et al., 2016). Although, in chronic SCI SSEP or SCEP could show improved function but the cord pathology is worsening. Thus, there is no clear agreement between these SCEP and SSEP monitoring for the cord pathology in long-term cases (see Mesbah et al., 2017; Stålberg et al., 1998). We further explored the relationship between SCEP and acute cord pathology to establish its prognostic values in a rat model of SCI (Sharma, Winkler, Stålberg, Olsson, & Dey, 1991; Winkler, Sharma, Stålberg, & Olsson, 1993). We have carried out a series of the experiments in which SCEP changes immediately after injury is correlated with later development of edema formation and cord pathology (see Mesbah et al., 2017). Furthermore, if we investigated further whether neuroprotective drugs when applied either immediately after or before injury could reflect SCEP changes after SCI with later development of cord pathology or edema formation (Mesbah et al., 2017; Sharma et al., 1991; Stålberg et al., 1998; Winkler et al., 1993).

2. Early SCEP changes could predict later development of spinal cord edema

Our observations showed that a focal SCI made on the right dorsal horn of the T10–11 segments (2 mm deep and 5 mm long) resulted in an immediate loss of SCEP amplitude as recorded from epidural microelectrodes placed above and below the lesion site after stimulation of right sciatic nerve (Mesbah et al., 2017; Sharma et al., 1991). This loss of amplitude was seen on the rostral recordings where the caudal recordings the SCEP amplitude even increased (Mesbah et al., 2017; Sharma et al., 1991, Fig. 1). Over the time the rostral SCEP amplitude partially recovers but the latencies on the rostral SCEP is increasing progressively with time (see Mesbah et al., 2017; Sharma et al., 1991) (Fig. 1). However, the caudal recording SCEP amplitude has increased further but the latency was also slowly increased over the 5h period (Sharma et al., 1991). A loss of SCEP amplitude over the rostral electrode suggests that the impulses are not going through above the lesion site (see Sharma et al., 1991). Increased SCEP amplitude below the lesion site suggests that the injury or cut may have damaged the inhibitory input resulting in loss of inhibitory potential causing higher amplitude below the lesion site (see Mesbah et al., 2017; Sharma et al., 1991).

In such a scenario when edema and cord pathology was measured after 5h SCI, the magnitude of edema formation correlated very well with the loss of rostral SCEP amplitude seen within 0–2min after SCI (Sharma et al., 1991). This suggests that electrical activity above the lesion site immediately after injury could be a result of both electrical network disturbances particularly loss of excitatory potentials and/or neurochemical mediators that are released quickly after injury (Mesbah et al., 2017; Sharma et al., 1991; Stålberg et al., 1998; Winkler et al., 1993). To further prove the role of neurochemicals in the spinal cord conduction we pretreated animals with a serotonin synthesis inhibitor *p*-Chlorophenylalanine (p-CPA) that inhibits profoundly the endogenous serotonin synthesis (Sharma et al., 1991). Serotonin is a well-known neurochemical mediator and regulates various functions in the spinal cord (Akiyama et al., 2017; Ganzer, Beringer, Shumsky, Nwaobasi, & Moxon, 2018; Kalinina, Zaitsev, & Vesselkin, 2018; Ma & Dun, 1986; Winkler et al., 1993). When SCI was made in these endogenous serotonin depleted animals no reduction in rostral SCEP amplitude was seen after injury (Sharma et al., 1991). Thus, rostral SCEP did not lose amplitude but continue to remain high even after injury. This suggests that trauma induced release of serotonin could also participate in the spinal cord conduction (Akiyama et al., 2017; Ganzer et al., 2018; Kalinina et al., 2018; Ma & Dun, 1986; Sharma et al., 1991; Winkler et al., 1993). In the drug treated animals SCI induced edema and cord pathology after 5h was also absent (Mesbah et al., 2017; Sharma et al., 1991).

When we pretreated the rats with a serotonin 5-HT2 receptor blocker cyproheptadine without interfering the serotonin synthesis or release after injury, the rostral SCEP amplitude not only lost its negativity but results

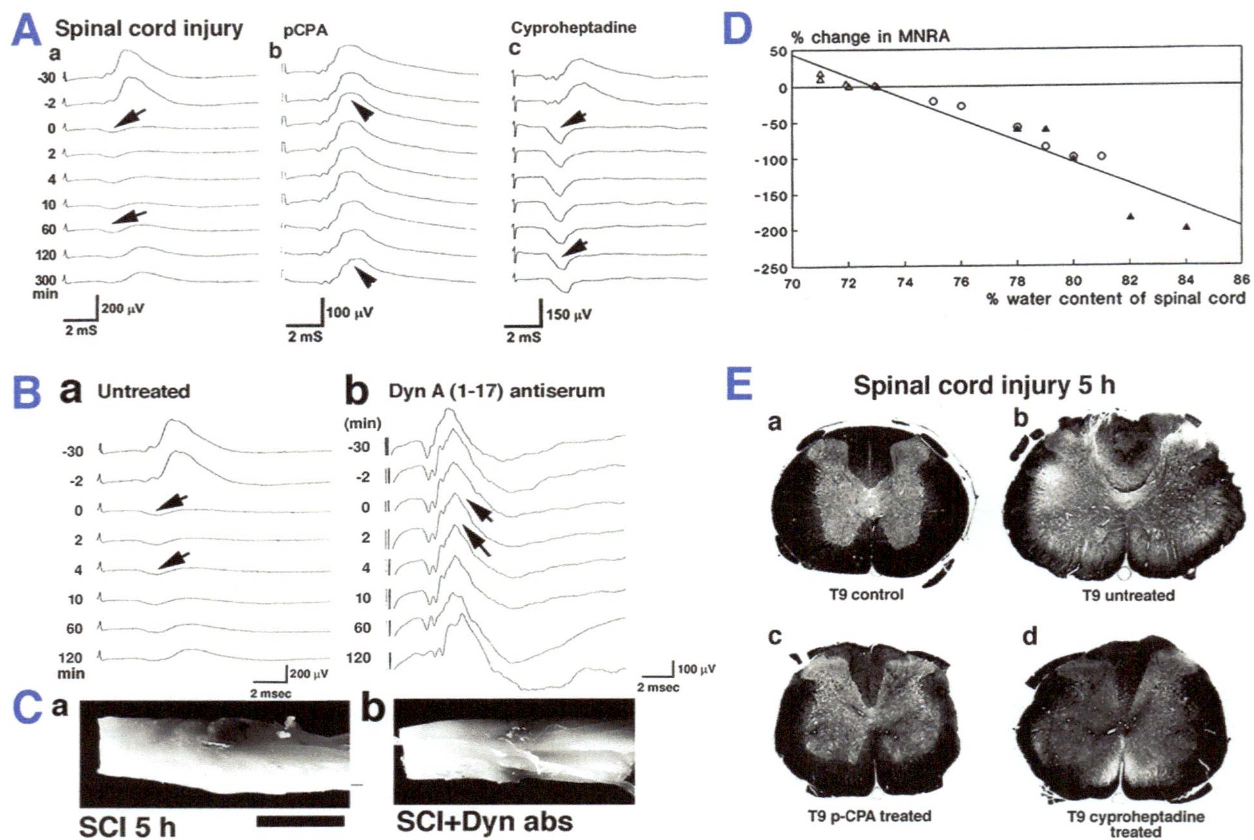

Fig. 1 See figure legend on opposite page.

in positive potential with increase in latencies over the 5 h period (Mesbah et al., 2017; Sharma et al., 1991). Measurement of cord pathology and edema showed higher accumulation of water content in this drug treated rat 5 h after SCI (Sharma et al., 1991). These observations confirm that early changes in the SCEP particularly above the lesion could have some prognostic value in later development of cord pathology (see Mesbah et al., 2017; Sharma et al., 1991; Stålberg et al., 1998). After that we tested our hypothesis using antibodies of neuronal nitric oxide synthase (nNOS) or antibodies to dynorphin A (1–17) topically applied over the traumatized spinal cord after 2 min injury (Feng et al., 2018; Niu et al., 2019; Sharma, 2004; Sharma & Westman, 2004; Sharma & Winkler, 2002; Winkler et al., 1998, 2002). We similarly found that rostral SCEP amplitude did not lost and maintained good negativity, but latency continue to rise over the 5 h period (Winkler et al., 2002; Winkler, Sharma, Stålberg, Badgaiyan, et al., 1998). These antibodies are supposed to block the nNOS upregulation after cord or dynorphin release that is harmful to the cord pathology. Spinal cord edema

Fig. 1 Spinal cord evoked potentials (SCEP) on Rostral recordings in untreated spinal cord injury (SCI) (A) (a) and their modification with p-CPA (b) and cyproheptadine (c) treatment. Loss of SCEP negative potential immediately after SCI showed significantly decline in untreated group (a) whereas p-CPA treatment completely prevented this decline after SCI (b). On the other hand cyproheptadine showed much adverse effects on SCEP (c) showing even development of positive potentials (arrows) after SCI. (B) Topical application of antibodies to dynorphin A (1–17) also prevented loss of SCEP amplitude (b) after SCI as compared to untreated SCI (a). (C) SCI 5 h shows profound edema and spinal cord damage (a), a feature not observed in antibody treated spinal cord (b) at 5 h SCI. (D) Tight correlation ($R^2 = 0.98$, $P < 0.001$) was seen between changes in SCEP negative potential after 2 min of SCI and development of spinal cord edema 5 h after injury. (E) Low power light micrograph showing edema and sponginess in untreated SCI 5 h (b) as compared to normal control spinal cord (a). p-CPA pretreatment significantly attenuated edema and expansion of the cord after 5 h SCI (c) whereas, cyproheptadine treatment did not offer any neuroprotection to the cord after 5 h (d). *Data modified from Sharma, H.S., Winkler, T., Stålberg, E., Olsson, Y., & Dey, P.K. (1991). Evaluation of traumatic spinal cord edema using evoked potentials recorded from the spinal epidural space. An experimental study in the rat.* Journal of the Neurological Sciences ***102*** *(2), 150–162; Sharma, H.S. & Winkler, T. (2002). Assessment of spinal cord pathology following trauma using early changes in the spinal cord evoked potentials: A pharmacological and morphological study in the rat.* Muscle & Nerve ***11*** *(Suppl.), S83–S91; Sharma, H.S. (2004). Pathophysiology of the blood-spinal cord barrier in traumatic injury. In: H.S. Sharma and J. Westman (Eds.),* The blood-spinal cord and brain barriers in health and disease, *Elsevier Academic Press, San Diego, pp. 437–518.*

and cord pathology was not evident in these antibodies treated traumatized rats after 5 h (Winkler et al., 2002; Winkler, Sharma, Stålberg, Badgaiyan, et al., 1998). These observations further confirm that early loss of SCEP negative potentials after injury could be a good prognostic factor in predicting later cord pathology (see Mesbah et al., 2017; Niu et al., 2019; Sharma, 2004; Sharma & Westman, 2004; Sharma & Winkler, 2002). It appears that these neurochemicals or antibodies applied over the traumatized cord help preserve SCEP potentials above the lesion (Mesbah et al., 2017; Sharma & Winkler, 2002; Winkler et al., 2002; Winkler, Sharma, Stålberg, Badgaiyan, et al., 1998). Thus, bridging the gap of spinal cord incision or tissue loss with good electric conductor such as titanium or titanium plate with out without nanowires grown over them could maintain good nerve conduction after SCI. This is also possible that apart from titanium or titanium plates implant good neuroprotective agents either applied over the spinal cord or administered intravenously could further potentiate spinal cord conduction and delay pathological processes within the cord, such as edema formation.

3. DL-3-*n*-butylphthalide is powerful neuroprotective agent

DL-3-*n*-butylphthalide (NBP) is synthetic drug that is most powerful antioxidant and inhibitor of apoptotic cell death (Feng et al., 2018; Niu et al., 2019). The drug is used in China since ages to treat well-being of people, mood elevation and to reduce hypertension and other mental diseases (see Abdoulaye & Guo, 2016). Clinical trials in stroke and Alzheimer's disease showed positive results in patients as compared to placebo (Cui et al., 2013; Jia et al., 2016). Our previous experiments show that NBP alone when given 30 min to 2 h after concussive head injury (CHI) showed remarkable neuroprotection after 8–12 h trauma (Feng et al., 2018; Niu et al., 2019). When NBP was administered through TiO_2-nanowired delivery 2–4 after CHI the nanodrug showed more superior neuroprotection and also this effect is prolonged to 24 and 48 h after the primary insult (Feng et al., 2018; Niu et al., 2019). This suggests that nanodelivery of NBP has superior neuroprotective ability in CHI. Few studies show that NBP is also neuroprotective in SCI and reduces blood-spinal cord barrier (BSCB) disruption together with functional improvements (He, Zhou, Lin, et al., 2017; He, Zhou, Huang, et al., 2017; Wang et al., 2018; Zheng et al., 2017). Thus, it would be interesting to evaluate whether nanodelivery of NBP in our model of SCI either applied topically or given intravenously could further potentiate the spinal cord neuroprotection (Tables 1 and 2).

Table 1 Effect of titanium implant on spinal cord injury (SCI) induced changes in the spinal cord evoked potentials (SCEP) and their modification with DL-3-*n*-butylphthalide (NBP) administration.

	Spinal cord evoked potentials (SCEP)									
	Amplitude (μV)					Latency (ms %)				
Types of expt.	−2 min	0 min	2 min	60 min	120 min	−2 min	0 min	2 min	60 min	120 min
SCI	118 ± 4	36 ± 4*	34 ± 5*	40 ± 4*	45 ± 6*	98 ± 2	96 ± 4	99 ± 3	110 ± 4*	119 ± 6*
SCI + Ti bare plate	104 ± 3	38 ± 3*	36 ± 4*	38 ± 6*	48 ± 8*	102 ± 3	98 ± 4	105 ± 6	108 ± 4*	118 ± 5*
SCI + Ti plate with TiO_2 NW	120 ± 5	44 ± 3*	48 ± 3*	50 ± 5*	52 ± 4*	104 ± 4	106 ± 7	110 ± 4*	114 ± 6*	123 ± 6*
SCI + Ti plate with TiO_2 NW + DL-NBP topical	116 ± 4	48 ± 3*	54 ± 5*	59 ± 5*	64 ± 5*	106 ± 3	100 ± 3	102 ± 5	118 ± 5*	128 ± 6*
SCI + Ti plate with TiO_2 NW + NW-DL-NBP topical	104 ± 3	62 ± 5*§	68 ± 3*§	74 ± 5*§	79 ± 3*§	102 ± 3	98 ± 3	110 ± 4*	118 ± 6*	128 ± 6*
SCI + Ti plate with TiO_2 NW + DL-NBP i.v.	102 ± 4	54 ± 3*	56 ± 2*	58 ± 6*	63 ± 4*	100 ± 2	102 ± 4	108 ± 6*	118 ± 4*	126 ± 6*
SCI + Ti plate with TiO_2 NW + NW-DL-NBP i.v.	104 ± 5	53 ± 4*	63 ± 5*	67 ± 5*	72 ± 6*	104 ± 4	104 ± 3	106 ± 3	112 ± 4*	122 ± 5*

Values are mean ± SD of 5–6 rats at each point. Topical DL-NBP or NW-DL-NBO 200 μL; intravenous DL-NBP or NW-DL-NBP 40 mg/kg, i.v. for details see text. *$P < 0.05$ from SCI control, §$P < 0.05$ from DL-NBP, ANOVA followed by Dunnett's test for multiple group comparison from one control.

Table 2 Effect of titanium implant on spinal cord injury (SCI) induced changes in the spinal cord evoked potentials (SCEP) and their modification with DL-3-*n*-butylphthalide (NBP) administration.

	Blood-spinal cord barrier (BSCB)			Spinal cord edema		
	Evans blue mg %			Spinal cord water content %		
	T9	T10–11	T12	T9	T10–11	T12
Types of expt.	**Rostral**	**Injured**	**Caudal**	**Rostral**	**Injured**	**Caudal**
SCI	1.56 ± 0.12	2.34 ± 0.32	1.89 ± 0.21	67.34 ± 0.22	68.34 ± 0.23	68.67 ± 0.34
SCI + Ti bare plate	1.34 ± 0.24	2.12 ± 0.14	1.65 ± 0.21	66.89 ± 0.27	67.94 ± 0.33	67.87 ± 0.23
SCI + Ti plate with TiO_2 NW	1.24 ± 0.44	1.98 ± 0.36	1.59 ± 0.46	66.16 ± 0.53	67.08 ± 0.63	67.05 ± 0.74
SCI + Ti plate with TiO_2 NW + DL-NBP topical	1.23 ± 0.32	1.89 ± 0.78	1.63 ± 0.21	66.05 ± 0.24	66.96 ± 0.52	66.88 ± 0.66
SCI + Ti plate with TiO_2 NW + NW-DL-NBP topical	1.18 ± 0.23*§	1.69 ± 0.20*§	1.39 ± 0.18*§	65.45 ± 0.34*§	65.79 ± 0.33*§	65.50 ± 0.41*§
SCI + Ti plate with TiO_2 NW + DL-NBP i.v.	1.43 ± 0.19	2.04 ± 0.32	1.87 ± 0.21	66.34 ± 0.21*	67.98 ± 0.65	67.34 ± 0.21*
SCI + Ti plate with TiO_2 NW + NW-DL-NBP i.v.	1.22 ± 0.21*	2.05 ± 0.21	1.53 ± 0.12*	66.13 ± 0.18*	66-97 ± 0.43*	67.01 ± 0.11

Values are mean ± SD of 5–6 rats at each point. Topical DL-NBP or NW-DL-NBO 200 μL; intravenous DL-NBP or NW-DL-NBP 40 mg/kg, i.v. for details see text. *$P < 0.05$ from SCI control, §$P < 0.05$ from DL-NBP, ANOVA followed by Dunnett's test for multiple group comparison from one control.

4. Combination of titanium implant and nanodelivery of NBP in spinal cord injury

We wanted to explore whether titanium implant alone could influence SCEP changes on rostral recording together with edema development after SCI. Furthermore, if the titanium implant on which TiO_2-naowires are grown are soaked in NBP solution and applied over the injured spinal cord could have better therapeutic effects with regard to cord pathology and edema development.

5. Methodological consideration

5.1 Animals

Biological experiments on spinal cord injury (SCI) was made on Male Wistar rats housed at controlled room temperature (21 ± 1 °C) with 12h dark and 12h light cycle and food and tap water were supplied ad libitum before the experiments. All the experiments were carried out according to NIH Guidelines and Care of experimental animals and all procedures are approved by the local Institutional Ethics Committee.

5.2 Spinal cord injury

Spinal cord injury (SCI) was inflicted in Equithesin anesthetized rats by making a longitudinal incision into the right dorsal horn of the T10–11 segments and the animal were allowed to survive 5h after injury (Sharma, Nyberg, & Olsson, 1992; Sharma & Olsson, 1990; Sharma, Olsson, & Cervós-Navarro, 1993; Sharma, Olsson, & Dey, 1990; Sharma, Olsson, Nyberg, & Dey, 1993). The dimensions of the incision were 2mm deep and 5mm long over the right dorsal horn (Sharma et al., 1990, 1992; Sharma & Olsson, 1990). Control animals were only laminectomized and no injury was performed.

5.3 Titanium implant

In separate group of rats, titanium implants (2mm wide and 5mm long) were placed over the traumatized animals either alone or with TiO_2-nanowires grown over the surface. The characteristics of titanium implant are described below.

6. Validity of nanomaterials for better conduction

Different methods have been developed to synthesize TiO_2 nanowires. One of the most known methods is the substrate or templet-based method in which the desired nanostructures can be grown or etched from the nonporous structure of the substrate (Dupuis, Jodin, & Rouvière, 2006; Martin, 1994). Usually this method can be performed with the help of a strong oxidizer such as H_2O_2 or NaOH. However, many synthesis techniques such as sol–gel deposition (Lakshmi, Patrissi, & Martin, 1997; Lei et al., 2001), electrochemically induced sol–gel method (Miao et al., 2002), sol–gel electrophoretic deposition (Lin, Wu, Yuan, Xie, & Zhang, 2003), and anodic oxidative hydrolysis method (Lei, Zhang, & Fan, 2001) were developed base on the same concept to fabricate desired nanostructures on surface of various templets materials.

Hydrothermal method, a non-templet method, has been also used to fabricate various TiO_2 nanostructures such as nanowire, nanotubes, nanoribbons and nanobelts from TiO_2 nanopowder (Kasuga, Hiramatsu, Hoson, Sekino, & Niihara, 1998; Kasuga, Hiramatsu, Hoson, Sekino, & Niihara, 1999; Morgado et al., 2007; Yuan & Su, 2004). One of the methods that have been used to increase the durability of implants is by etching nanowires on the surface. This method used extensively to improve the performance and the overall durability and strength of the medical implants and equipment. In addition, nanostructure etching on the surface of implant can improve the biocompatibility and tribological characteristics of implant surfaces (Biggs, Richards, & Dalby, 2010; Dalby et al., 2006; Dinan, Gallego-Perez, Lee, Hansford, & Akbar, 2013; Goriainov et al., 2018). Despite of all the favorable properties of the titanium implant for the medical devices, it has been reported that Ti implant may show a cohesive failure or delamination even after coating with a thin nitride layer. These flaws might be duo to a defect in the surface preparation or unoptimized coating process (van Hove, Sierevelt, van Royen, & Nolte, 2015). Therefore, present study intends to minimize the defect associated with the use of a bare metal surface by etching a well-attached nanostructures on the surface of the Ti plate and assess the associated effects on cell regeneration and growth.

The versatile properties such as the semiconductivity, piezoelectric property, spontaneous polarization, mechanical, and thermal stability of titanate nanowires made these materials very important in many applications such as solid-state lighting, electrical devices manufacturing, skin care products, and water purification (Choi, Choi, Yang, Kim, & Yu, 2016; Fateh, Dillert, & Bahnemann, 2014; Othman, Amat, Ahmad, & Rajan, 2014; Wu et al., 2017). Recently, nanoscale generators have been developed from TiO_2 nanoparticles and these nanostructures revealed a unique property by producing a constant and repeatable electrical current (Dallacasa, 2012). Therefore, TiO_2 nanostructure might be a promising addition to the biomedical implants due to their piezoelectric property. Piezoelectric materials have shown a great biological effect when they use in bone implant to enhance cells growth. Basically, these materials can interpret the environmental signals into biological acceptable signals which might positively alter one of the cell pathways to enhance the growth (Marino & Becker, 1970).

7. Technological consideration

7.1 Materials

Titanium foils (0.1 mm thickness, purity: 99.6%, Advent Materials); Sodium Hydroxide pallet (reagent grade, Carolina), concentrated Ammonium Hydroxide solution (NH_4OH 28–30%, EMD); Acetone solvent solution (ASTM International); Titanium Dioxide (TiO_2) nanopowder (Degussa P25); Deionized water prepared by ion exchange in or lab.

7.2 Methods

7.2.1 Etching of titanium nanowires on titanium plate

Titanium foils with a size of (1 mm × 10 mm) has been precisely cut and flatten. Then the acquired pieces have been washed and sonicated with acetone for 15 min. The clean Ti foils then dried in the oven for 2 h at 45 °C. For the nanowires etching, hydrothermal process has been conducted in single alkali media (1 M of sodium hydroxide). Ten Ti foils were used with NaOH and another 10 were kept untreated to serve as control. Each Ti foil were immersed in 1 M of NaOH and sealed in a stainless-steel Teflon linear then heated in the oven for 6 h at 240 °C. The resulting foils were then washed with DI water several times until the surface pH reaches 7.2. This method has been achieved according to Kim, Miyaji, Kokubo, Nishiguchi, and Nakamura (1999) with some modifications.

7.2.2 Synthesis of titanium nanowires in suspension

A of 0.375 g of Titanium Dioxide nanopowder (Degussa, P25) was mixed with 50 mL of 10 M NaOH solution and left on the magnetic stirrer for 24 h. Then the obtained solution was sonicated for 15 min before the thermal treatment. The resulting solution was then transferred to a 100 m stainless steel teflon linear and sealed well before the cooking. The sealed teflon linear was then cooked in the oven at 240 °C for 72 h. After the cooking was done, the obtained suspension was washed with DI water several times alternating between sonication and vortex for 5 min between the washes. Washing process was repeated for about 20 times until the resulting pasty material get neutralized (pH 7.2). This process was followed to prepare the sodium version of titanium dioxide nanowires (TiO_2 nw) (Nor et al., 2012).

7.3 Characterization

7.3.1 X-ray diffraction analysis (XRD)

XRD analysis was conducted to evaluate the crystal structure of both control and treated samples. Fig. 2 represents the XRD pattern for all subjected samples. X-ray diffraction patterns were recorded from 5° to 80° (2θ) with a step width of 0.02, using a Rigaku Desktop X-Ray Diffractometer. TiO_2 anatase phase can be indicated by the diffraction peaks at 25.2°, 37.8°, 48.2° and 55.2° according to JCPDS 83-2243 and JCPDS no 84-1286, while the sharp peaks at 38.4°, 40.2°, 53.0° and 70.6° might be attributed to the background of metallic Ti (Wu et al., 2009). Sodium titanate nanowires diffraction peaks were shown on $2\theta = 9.788$ and/or 27.858 can be assigned to $Na_2Ti_2O_5 \cdot {}_2O$ (JCPDS No. 47-0124) (Fig. 2B and C). However, these peaks are very similar to the peaks ascribed to the plane structure of the $H_2Ti_2O_5 \cdot H_2O$ which are well indicated in previous study (Tsai & Teng, 2006). Present results suggest that the resulted nanowires obtained by the hydrothermal reaction with NaOH are $H_2Ti_2O_5 \cdot H_2O$ phase nanowires duo to the ion exchange processing with DI water after the hydrothermal treatment. The $H_2Ti_2O_5 \cdot H_2O$ has been assigned to anatase phase of TiO_2. This result is in well agreement with previous study which stated that the ion exchange after the hydrothermal synthesis of Na-TiO_2 nanowires resulted in the formation of H-Titanate nanowires (Chu, Qin, & Yang, 2015).

7.3.2 Scanning electron microscopy (SEM) imaging

Etching of nanowires on Ti plate by using hydrogen peroxide has been reported and used for biomedical application such as surface modification of implants and medical instruments. In most reported studies, nanowires fabricated on the surface of Ti plate by using direct oxidation of the surface with hydrogen peroxide and low concentration of alkali (Chen & Mao, 2007). In this study, nanowires were fabricated on the surface of Ti plate by using strong pure alkali media (NaOH) to obtain sodium titanate

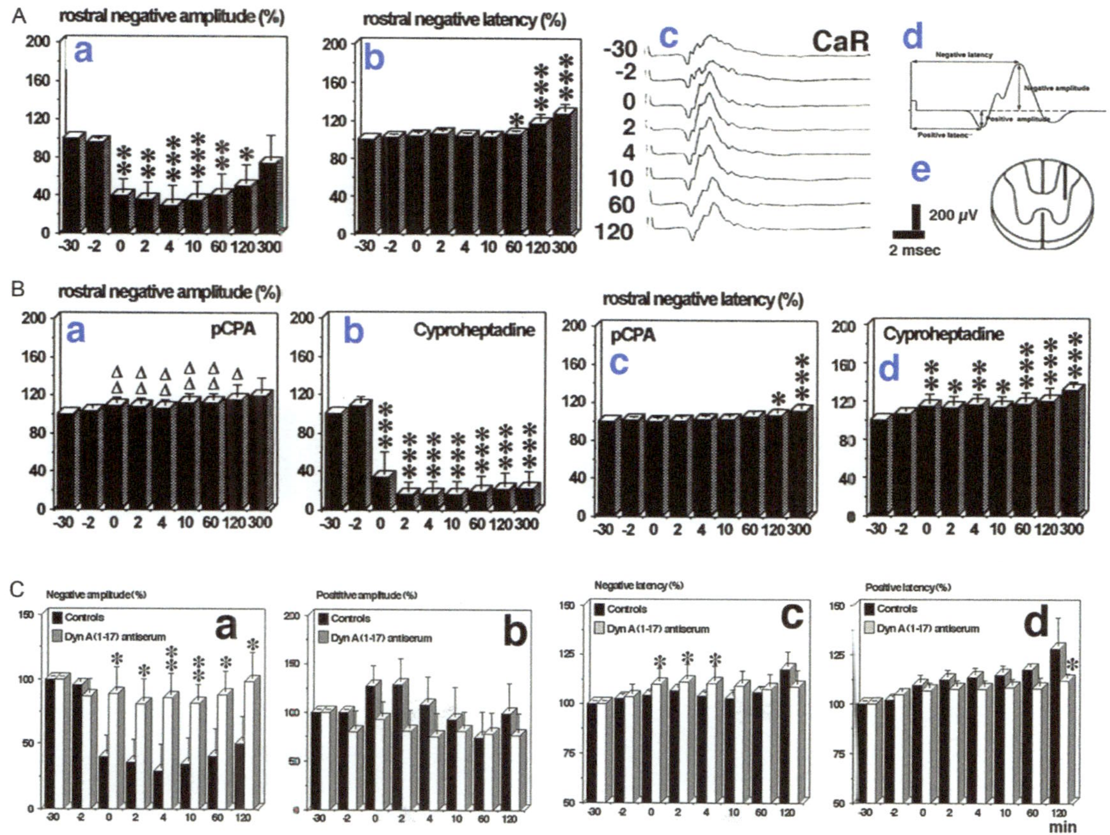

Fig. 2 See figure legend on next page.

nanowires on plate. Also, titanate nanowires alone were synthesized using the substrate TiO_2 nanopowder (Degussa P25) as prescribed previously.

SEM results showed a couple differences between the nanowires etched on the titanium plate and the nanowires produced based on the TiO_2 powder. In Fig. 3a and b, SEM images showed the difference between the bare Ti plate topology and the Ti plate with nanowires on the surface. It is obvious that the necked Ti has a surface lacked to the topological structures that are required to improve the attachment of the tissue on the surface (van Hove et al., 2015). However, present results show that fabricating TiO_2 nanowires on the surface may be more efficient since the nanowires are well immersed in the surface. Furthermore, by etching nanostructure on the surface will increase the surface area of the implant which in turn may increase the catalytic activity of total surface area (Yang et al., 2008). Interestingly, SEM images also showed a quite difference between the morphology of the nanowires gown on the surface of Ti plate (Fig. 3b and c) and the titanium nanowires alone (Fig. 3d, e, and f). Titanate nanowires fabricated by soaking the Ti plate in 1 M NaOH formed a nest or plant root-like structures on the surface of the titanium plate. On the other hand, nanowires synthesized based on TiO_2 nanopowder revealed a sticks-like structures that crossed together to form a mesh.

Fig. 2 Spinal cord evoked potential amplitude (A) (a) and latency (b) changes after untreated spinal cord injury (SCI). (c) SCEP recording on caudal electrode in untreated SCI showed no loss of SCEP negative amplitude rather there was an increase in SCEP amplitude after SCI (c). Method of SCEP negative amplitude calculations are shown in (A). (d) A model of SCI showing lesion of the right dossal horn is shown in (e). (B) Changes in SCEP amplitude in SCI treated with p-CPA (a) and cyproheptadine (b) as well as rostral latency changes in p-CPA (c) and cyproheptadine (d) after SCI. (C) Mean SCEP negative amplitude (a), positive amplitude (b), negative latency (c) and positive latency (d) in control and dynorphin antibody treated SCI. $^{***}P < 0.001$, $^{**}P < 0.01$, $^{*}P < 0.05$ from corresponding controls. Student's unpaired t-test. *Data modified from Sharma, H.S., Winkler, T., Stålberg, E., Olsson, Y., & Dey, P.K. (1991). Evaluation of traumatic spinal cord edema using evoked potentials recorded from the spinal epidural space. An experimental study in the rat.* Journal of the Neurological Sciences ***102*** *(2), 150–162; Sharma, H.S. & Winkler, T. (2002). Assessment of spinal cord pathology following trauma using early changes in the spinal cord evoked potentials: A pharmacological and morphological study in the rat.* Muscle & Nerve ***11*** *(Suppl.), S83–S91; Sharma, H.S. (2004). Pathophysiology of the blood-spinal cord barrier in traumatic injury. In: H.S. Sharma and J. Westman (Eds.),* The blood-spinal cord and brain barriers in health and disease, *Elsevier Academic Press, San Diego, pp. 437–518.*

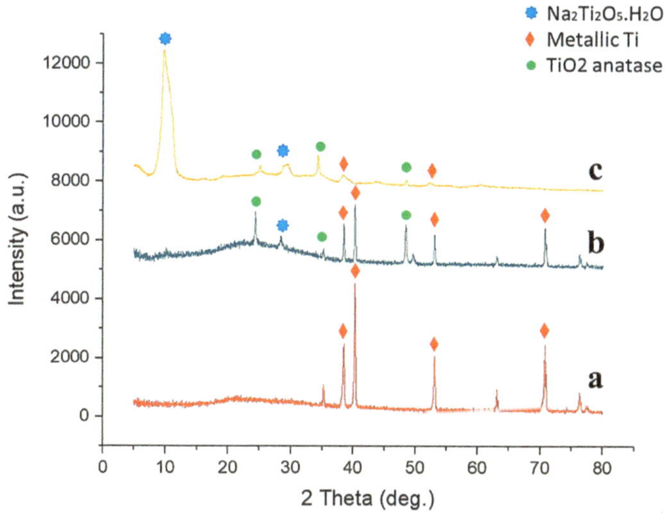

Fig. 3 XRD pattern for the prepared samples as indicated; (a) is a bare titanium plate (control sample); (b) the crystallographic pattern of the sodium version of titanate nanowires (TiO_2) etched on titanium plate; and (c) sodium version of TiO_2 nanowires synthesized by the hydrothermal method from the TiO_2 powder.

Moreover, nanowires etched on the surface has a thinner diameter of about 100nm compared to the 100–1000nm of the titanate nanowires alone. However, the length of both types of nanowires seems to be similar (15–20μm). Therefore, the final products (on plate nw vs nw alone) are expected to show a totally different behaviors for biological uses since the nanoscale morphology is different. It has been stated that the variance in morphology results in different quantum confinement effects as a result of a change in electron transport (Roduner, 2006). In addition, surface to volume ratio of nanostructure may have a huge impact on the catalytic activity between materials (Valden, Lai, & Goodman, 1998). The increase in this ratio leads to increase in the surface reaction sites which in turn improves material catalytic activity and may also stabilize the material (Yang et al., 2008). Therefore, the applied improvement on the surface of Ti implant is required.

7.3.3 Energy dispersed X-ray microanalysis

Energy dispersive X-ray spectroscopy (EDX) analysis has been performed using (Bruker AXS Microanalysis GmbH Berlin, Germany) EDX system.

Present study showed that Na elements were shown only in samples with nanowires (Fig. 3b and c). This finding my duo to the high content of the Na in the samples which diffused in the lattice of the nanowires structure during the hydrothermal process with NaOH. Also, this amount has not been reduced by the ion exchange processing with DI water. In order to preserve the morphology of the obtained nanostructures, ion exchange with HCl has been avoided in present study since processing with HCl might destroy the structure of the fabricated nanowires which in turn might change the dimension and correlated activity of these nanostructures. Previous study stated that the length of the fabricated nanostructures by the hydrothermal method can be significantly shorten after ion exchange with HCl (Bela, Wong, & Ho, 2010). Finally, carbone C and oxygen O concentrations were both increased after nanowires synthesis compared to the bare Ti plate sample which might be duo to hydrothermal processing and the formation of TiO_2 nanowires.

7.4 SCEP recording in SCI

For SCEP recordings, two epidural electrodes were placed above and below the lesion site on the right side over the dorsal horn. The active surface of the electrodes was lying over the T9 (rostral) and T12 (caudal) segments (see Mesbah et al., 2017; Sharma et al., 1991; Sharma & Winkler, 2002; Winkler et al., 1993, 2002; Winkler, Sharma, Stålberg, Badgaiyan, et al., 1998). The lesion was made on the right dorsal horn of the T10–11 segments as described earlier (Sharma et al., 1991). Right sciatic and sural nerves were stimulated maximally and the SCEP were obtained as described earlier (see Sharma et al., 1991). The mean negative amplitude and latencies of rostral and caudal SCEP were calculated for statistical significance (Mesbah et al., 2017; Sharma et al., 1991).

7.5 DL-NBP administration

DL-NBP either alone or TiO_2 nanowired-NBP (concentration 10 mg/ml solution) were either applied topically over the traumatized spinal cord (200 μL) immediate after injury followed by 1 h and 2 h posttrauma with or without titanium implants over the injured tissue. In another sent of experiments, 60 mg/kg normal NBP or 40 mL/kg TiO_2-NBP was administered immediately after injury followed by 1 and 2 h post trauma with or without titanium implants.

7.6 Motor function after SCI

The Tarlov scale was used to assess motor function of the spinal cord injured animals as described earlier (Sharma et al., 2007, 2009).

7.6.1 Blood-spinal cord barrier permeability

The blood-spinal cord barrier (BSCB) was evaluated using Evans blue albumin. The dye was administered (2% of 3mL/kg, i.v.) 10min before termination of the experiment (Sharma, 2007, 2010; Sharma et al., 1998; Sharma, Olsson, & Westman, 1995). After the end of the experiments, the intravascular tracer was washed out with in situ perfusion with cold physiological saline at 100Torr (Sharma et al., 1990, 1992; Sharma & Olsson, 1990; Sharma, Olsson, & Cervós-Navarro, 1993; Sharma, Olsson, Nyberg, & Dey, 1993). After that the spinal cord tissue was dissected out and spread of dye was examined in a blinded fashion by at least two investigators and recorded (Sharma & Olsson, 1990). The dye entered into the spinal cord tissue was extracted according to standard protocol and measured in spectrophotometer at 620nm (Sharma et al., 1990, 1992, 1992; Sharma & Olsson, 1990; Sharma, Olsson, & Cervós-Navarro, 1993; Sharma, Olsson, Nyberg, & Dey, 1993).

7.6.2 Spinal cord edema

After 5h, the spinal cord segments T9, T10–11 and T12 were dissected out and immediately weighed on a preweighed filter paper (Sharma & Olsson, 1990). After recording their wet weight, the samples were placed in an oven maintained at 90°C for 72h to evaporate all tissue water from the cord (Sharma et al., 1990, 1992; Sharma, Olsson, & Cervós-Navarro, 1993). When the dry weight of the samples became constant, water content was calculated from differences between wet and dry weight of the cord samples as described earlier (Sharma et al., 1990, 1992; Sharma & Olsson, 1990; Sharma, Olsson, & Cervós-Navarro, 1993; Sharma, Olsson, Nyberg, & Dey, 1993) (Fig. 4).

8. Our observations in SCI

We evaluated SCI pathology on the SCEP recordings, edema formation and the BSCB disturbances in all groups of SCI and controls. A brief summary of the findings is described below.

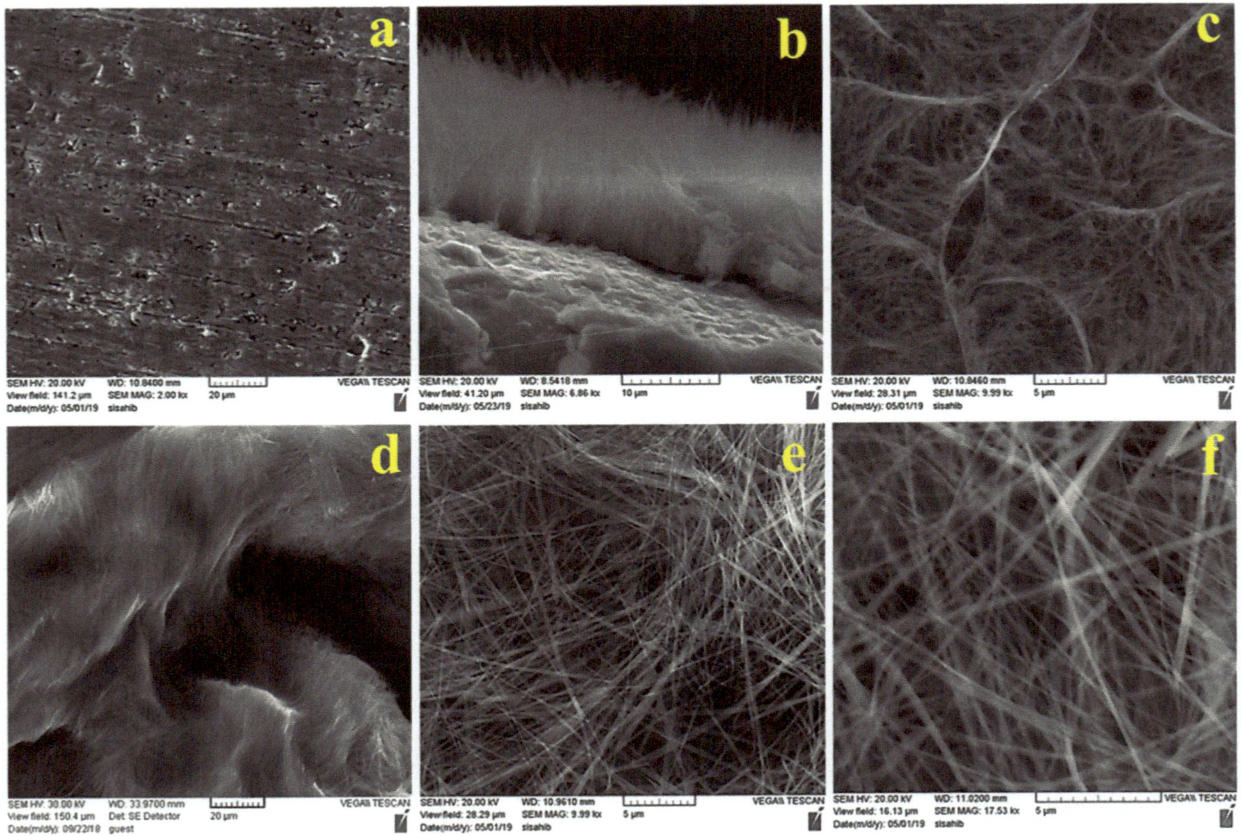

Fig. 4 SEM image of (a) bare Ti plate surface; (b and c) side and top view of the sodium version of Titanate nanowires (TiO_2) grown on the surface of Ti plate; (d, e, and f) sodium version of TiO_2 nanowires synthesized by the hydrothermal method from the TiO_2 powder at different magnifications.

8.1 Effect of titanium implants on SCI

Titanium implants over the injured cord resulted in only slight changes in SCEP rostral negative amplitude was recorded after stimulation of right tibial and sural nerves from the rostral recordings. Thus, there was not a 100% loss of negative amplitude and some negative amplitude with increased latency was seen over the 5h periods in this group. When titanium implants were placed with TiO_2 nanowires grown on the surface over the injured spinal cord some improvement in SCEP negative amplitude was seen as compared to titanium implants alone. However, this improvement is slight but significant after SCI and maintained over the 5h period. The latency continued to increase and was not significantly different from titanium implants alone.

In titanium implant traumatized rats, the BSCB breakdown to Evans blue showed only a minor decrease in the three cord segments. The most marked decrease in the BSCB breakdown to Evans blue was seen in the T9 segment above the lesion site. Whereas, injured segment T10–11 and segment below the lesion site T12 showed very little reduction in the BSCB breakdown. This is also true for titanium implants with TiO_2 nanowires over the spinal cord injured tissues. Control rats implanted with titanium and/or titanium plus TiO_2 nanowires did not show any significant alterations from sham group without any implant.

8.2 Effect of DL-NBP in SCI

Topical application of DL-NBP over the traumatic spinal cord (200μL) immediately after injury significantly preserved rostral SCEP negative amplitude over the 5h, however, latency of the negative amplitude continued to increase over 5h period. When TiO_2 nanowired DL-NBP was applied topically over the traumatized cord, SCEP negative amplitude was also preserved better than the normal NBP. However, the latency continued to increase although a bit slower that the NBP alone over the 5h period. Intravenous application of normal NBP 60mg/kg or TiO_2 nanowired NBP 40mg/kg, i.v. both are able to sustain loss of negative amplitude on the rostral recordings after injury. However, the degree of loss of mean negative amplitude is greater than those compared to topical application of the DL-NBP. The latency continues to increase over 5h period following SCI.

The BSCB breakdown and edema formation were slightly but significantly reduced by NBP application either through intravenous rote or its topical application. TiO_2 nanowired NBP has superior effects in reducing

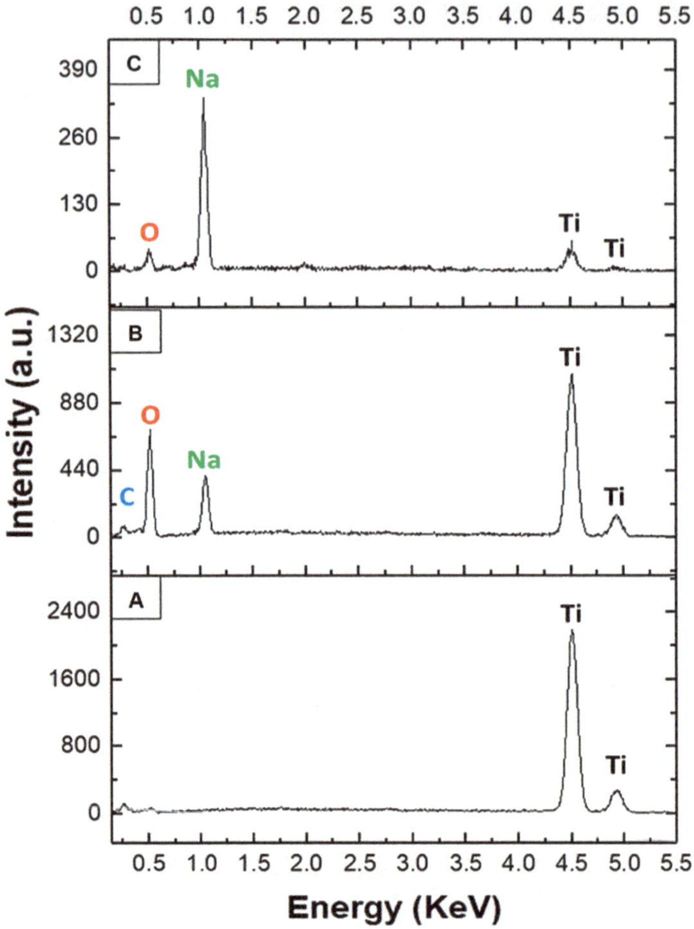

Fig. 5 EDX spectrum for (A) bare Ti plate; (B) sodium version of titanate nanowires (TiO_2) etched on Ti plate; (C) sodium version of TiO_2 nanowires synthesized by the hydrothermal method from the TiO_2 powder.

BSCB disruption to Evans blue and edema formation either given intravenously or applied topically as compared to identical application of normal NBP in SCI (Fig. 5).

8.3 Effect of DL-NBP and titanium implant on SCI

When titanium implant with or without TiO_2 nanowires were soaked in DL-NBP were applied over the injured spinal cord followed by topical application on 200 μL DL-NBP immediately after injury, the SCEP negative amplitudes on rostral recordings were significantly preserved (almost

70% or the pre-injury level) that was maintained somehow at the same level up to 5 h SCI. The latency continued to increase over the 5 h period. However, latencies of the SCEP negative amplitude continued to increase over the 5 h period. These effects were on SCEP negative amplitude on rostral recordings was further enhanced after SCI when TiO_2-NBP was applied (200 μL) with titanium implant containing TiO_2 nanowired placed over the injury site after soaking in TiO_2-nanowired NBP solution.

Intravenous administration of DL-NBP alone or TiO_2 nanowired showed much less neuroprotection on SCEP activity as compared to its topical application with titanium implant with or without TiO_2-nanowires grown over the titanium surface. The BSCB and edema formation showed most remarkable reduction in titanium implant with TiO_2 nanowires soaked in TiO_2 nanowired solution followed by its topical application over the spinal cord as compared to all other intravenous application of normal or TiO_2 nanowired NBP application.

9. Spinal cord motor functions after titanium implants and DL-NBP treatment

The Tarlov scale showed a close parallelism between SCEP negative amplitude preservation and improved motor function. Thus, maximum improvement in motor function was seen when titanium implants with TiO_2 nanowires soaked in nanowired NBP solution was applied over the injured spinal cord followed by topical application of the nanowired NBP. Other combination of intravenous administration of NBP with or without nanowires after titanium implant with or without nanowires coated showed only mild improvement in Tarlov scale after 5 h SCI.

10. Possible mechanisms of neuroprotection in SCI and functional significance

The novel findings of this investigation show that titanium implants with TiO_2 nanowired when soaked in nanowired DL-NBP solution followed by topical application of TiO_2 nanowired DL-NBP (200 μL) immediately after injury significantly attenuated loss of SCEP negative amplitude on rostral recordings after SCI that was more or less maintained until 5 h. This suggests that a combination of titanium implant with nanowires together with nanowired delivery of neuroprotective agent DL-NBP enhanced and maintained spinal cord conduction after injury that is continued until 5 h period.

On the other hand, titanium implants without TiO_2 nanowires over its surface when soaked in normal NBP following topical application of NBP (200 μL) after injury has much less preservation of SCEP negative amplitudes after SCI on rostral recordings. This suggests that to enhance superior spinal cord conduction with titanium implants associated with neuroprotective drug nanodelivery of NBP and nanowires titanium implants are needed. Topical application of nanowired DL-NBP alone is also able to enhance spinal cord conduction after injury but the magnitude and intensity of the effects are much lower. This suggests that apart from drug effects, titanium implants with TiO_2 nanowires significantly potentiated the nanowired NBP effects on spinal cord conduction.

The probable mechanisms by which nanowired coated titanium implants enhanced the spinal cord conduction effects of nanowired NBP are not well understood. However, it appears that nanowired titanium implants soaked in nanowired NBP with additional topical application of nanowired NBP may create a bond between titanium surface and nanowired material to extend neuronal connections leading to superior spinal cord conduction seen after SCI. We have used a longitudinal incision of the spinal cord dorsal horn limited to the lamina VIII of the spinal cord (see Mesbah et al., 2017; Sharma et al., 1991). This would leave spinal ventral horn completely untouched by the physical lesion. Although release of neurochemicals within the spinal cord and swelling of the cord could affects neuronal networks in the ventral spinal horn as well as described earlier (Sharma et al., 1990, 1992; Sharma & Olsson, 1990; Sharma, Olsson, & Cervós-Navarro, 1993; Sharma, Olsson, Nyberg, & Dey, 1993). However, application of nanowired NBP and combination of titanium implant with nanowires soaked in the same solution could rapidly penetrate the ventral spinal horn to alleviate swelling and preservation of the neuronal network. In addition, lesion of the dorsal horn where most of the negative potential is generated is adversely affected by release of several neurochemicals affecting spinal cord conduction (Mesbah et al., 2017; Sharma et al., 1991; Sharma & Winkler, 2002; Winkler et al., 2002; Winkler, Sharma, Stålberg, Badgaiyan, et al., 1998). It is quite likely that nanowired delivery of NBP together with titanium implant with TiO_2 nanowires soaked with the same drug could minimize or neutralize the adverse effects of the released neurochemicals on negative potential. As a result, SCEP negative potential is maximally preserved in this arrangement as compared to any other arrangement we have seen. It would be interesting if we could obtain similar enhancement of spinal cord conduction after SCI where a transverse lesion of the spinal cord is made, or a compression or contusion injury is inflicted on the whole spinal cord. This is a feature that requires additional investigation.

Intravenous administration of TiO_2 nanowired or normal NBP after titanium implants with or without TiO_2 nanowires did not induce greater neuroprotection in SCEP amplitude. This suggests that the concentration of drug is not enough at the site of injury after intravenous administration as compared to topical application. Obviously, the drug concentration reaching to the spinal cord after intravenous administration is much diluted as compared to the local application.

Topical application of either nanowired or normal NBP when applied over the injured cord with titanium implants without TiO_2 nanowires on the surface suggests that combination of nanowired NBP and nanowired titanium implants are better suitable for enhanced spinal cord conduction. Whether, combination of nanowired material on implants and nanodelivery of drug could create a local chemical or electrochemical bond over the injured spinal cord tissue enhancing spinal cord conduction after injury is unclear. Alternatively, nanowired drug and nanomaterial on p titanium implant may effectively neutralize the release of neurochemicals within the dorsal horn after injury to maintain negative potentials of the dorsal horn (Sharma et al., 1991; Sharma & Winkler, 2002; Winkler et al., 2002; Winkler, Sharma, Stålberg, Badgaiyan, et al., 1998). However, further investigation using biochemical approach to find out the neurochemical environment of the dorsal horn after titanium implant and topical application of drugs are needed to clarify these points.

Another key point emerged from this study showing that although spinal cord conduction is significantly enhanced after SCI in combination with titanium implants containing TiO_2 nanowires and topical application of nanowired NBP up to 5 h period. Whether this combination could also enhance or maintain the spinal cord conduction after days or weeks is a matter of further investigation. Also, long-term titanium implant with nanowires and their interaction with spinal cord tissue with regard to any cellular reactivity or neurotoxicity is another important aspect that should be investigated in greater details. Maintenance of spinal cord conduction on the rostral electrodes showed a close parallelism with improved Tarlov scale (see Mesbah et al., 2017; Sharma et al., 1991; Stålberg et al., 1998). This suggests that functional motor improvement is associated with better spinal cord conduction. It remains to be seen whether this improved motor function over 5 h period is sustainable to longer periods of time after SCI is the burning question at the moment. This is a feature that is currently being investigated in our laboratory.

A reduction in the BSCB disruption and edema formation seen after SCI also showed a close interrelationship with SCEP negative amplitude preservation on the rostral recordings (Sharma et al., 1991; Sharma & Winkler, 2002). This suggests that SCEP changes on the rostral recordings immediately after injury may have reliable prognostic values for cord pathology as described earlier (Mesbah et al., 2017; Sharma et al., 1991; Sharma & Winkler, 2002). SCI induces disruption of the BSCB either directly due to physical lesion or microhemorrhages as well as indirectly through release of several neurochemical mediators of BSCB permeability is the key factors in formation of vasogenic edema (Sharma et al., 1990, 1992; Sharma & Olsson, 1990; Sharma, Olsson, & Cervós-Navarro, 1993; Sharma, Olsson, Nyberg, & Dey, 1993). Several neurochemicals such as serotonin, prostaglandin, histamine, dynorphin, nitric oxide, carbon monoxide etc. are known to be involved in BSCB disruption after SCI (see Mesbah et al., 2017; Sharma et al., 1991; Sharma & Winkler, 2002). Thus, blockade of these neurochemicals and other agents stimulating oxidative stress such as NO and CO results in restoration of the BSCB function and reduction in spinal cord edema together with improved spinal cord conduction (see Sharma et al., 1995; Winkler et al., 2000, 2003; Winkler, Sharma, Stålberg, & Badgaiyan, 2000; Winkler, Sharma, Stålberg, Olsson, & Nyberg, 1994a, 1994b, 1995; Winkler, Sharma, Stålberg, & Westman, 1997). This suggests that neurochemical mediators are key players in maintaining spinal cord conduction and SCEP negative amplitude above the lesion site after SCI (Mesbah et al., 2017; Sharma et al., 1991, 1995; Sharma & Winkler, 2002; Winkler et al., 1994a, 1994b, 1995, 1997, 2002, 2003; Winkler, Sharma, Stålberg, & Badgaiyan, 2000; Winkler, Sharma, Stålberg, Badgaiyan, et al., 1998; Winkler, Sharma, Stålberg, Badgaiyan, Westman, et al., 2000).

That neurochemical mediators and neurotrophic factors are involved in spinal cord conduction is further evidenced from the findings that topical application of brain derived neurotrophic factors (BDNF), insulin like growth factor-1 (IGF-1) or growth hormone (GH) after SCI is able to maintain spinal cord conduction and reduction in BSCB disturbances and edema formation (see Sharma et al., 1995; Winkler et al., 1994a, 1994b, 1995, 1997, 2003; Winkler, Sharma, Stålberg, & Badgaiyan, 2000; Winkler, Sharma, Stålberg, Badgaiyan, Westman, et al., 2000). It appears that topical application of nanowired NBP together with nanowired titanium implants could reduce oxidative stress profoundly and neutralize the adverse effects of other neurochemicals resulting in improved SCEP activity and reduction in the BSCB disruption and edema formation. Thus, it would be interesting to examine whether nanowired delivery of BDNF, IGF-1 or GH topically together with nanowired titanium implants could induce superior spinal cord conduction and neuroprotection in SCI.

Increased BSCB permeability to proteins induces vasogenic edema formation (Sharma et al., 1990, 1992; Sharma & Olsson, 1990; Sharma, Olsson, & Cervós-Navarro, 1993; Sharma, Olsson, Nyberg, & Dey, 1993). Leakage of Evans blue after SCI represents the extravasation of serum protein complex into the spinal cord (Martin, 1994; Sharma, 2007, 2010; Sharma et al., 1990, 1992, 1995, 1998, 2007, 2009; Sharma & Olsson, 1990; Sharma, Olsson, & Cervós-Navarro, 1993; Sharma, Olsson, Nyberg, & Dey, 1993). One molecule of Evans blue when administered into the systemic circulation binds to about 12 molecules of serum albumin (Sharma et al., 1990, 1992, 1992; Sharma & Olsson, 1990; Sharma, Olsson, & Cervós-Navarro, 1993; Sharma, Olsson, Nyberg, & Dey, 1993). Thus, leakage of Evans blue in the spinal cord represents extravasation of dye-protein complex (MW 68 kD) associated with vasogenic edema formation (Martin, 1994; Sharma, 2007, 2010; Sharma et al., 1995, 1998, 2007, 2009).

11. Conclusion and future perspectives

In conclusion, our observations are the first to show that topical application of nanowired NBP in combination with TiO_2 nanowired titanium implant induced superior neuroprotection and spinal cord conduction after SCI. Since titanium implant with nanowires alone has some neuroprotective effects with improvement in spinal cord conduction after SCI, a possibility exists that such implants in future could open new vistas in patients to minimize cord pathology and sensory motor dysfunction with the use of suitable drugs, such as nanowired DL-NBP. In addition, long-term survival of SCI up to weeks and months are needed to understand the detailed effects of titanium implants with nanowires and topical delivery of nanowired drugs to enhance quality of life of SCI victims. Obviously, in such long-term cases, neurotoxicity of titanium implants requires additional investigation. These features are currently the prime goal of future investigations in SCI in our laboratory.

Acknowledgments

Dr. Seaab Sahib is supported by Research Fellowship at University of Arkansas Fayetteville AR by Department of Community Health; Middle Technical University; Wassit; Iraq, and The Higher Committee for Education Development in Iraq; Baghdad; Iraq. This investigation is partially supported by grants from Ministry of Science & Technology, People Republic of China, the Air Force Office of Scientific Research (EOARD, London, UK), and Air Force Material Command, USAF, under grant number FA8655-05-1-3065; the National Institutes of Health (R01 AG028679); Swedish Medical Research Council (Nr 320 & 2710-HSS), Göran Gustafsson Foundation, Stockholm, Sweden (HSS), Astra Zeneca, Mölndal, Sweden (HSS/AS), The University Grants Commission, New Delhi, India (HSS/AS), Ministry of Science & Technology, Govt. of India (HSS/AS), Indian Medical Research Council, New Delhi, India (HSS/AS) and India-EU Co-operation Program (AS/HSS) and Society for Neuroprotection and Neuroplasticity (SSNN), Romania. Technical assistance of Madeline Järild, Gunilla Tibling, Kerstin Flink and Ingmarie Olsson (Uppsala University) is highly appreciated with thanks. We thank Suraj Sharma, Uppsala, Sweden for computer and graphic support. The U.S. Government is authorized to reproduce and distribute reprints for Government purpose notwithstanding any copyright notation thereon. The views and conclusions contained herein are those of the authors and should not be interpreted as necessarily representing the official policies or endorsements, either expressed or implied, of the Air Force Office of Scientific Research or the U.S. Government.

References

Abdoulaye, I. A., & Guo, Y. J. (2016). A review of recent advances in neuroprotective potential of 3-N-butylphthalide and its derivatives. *BioMed Research International, 2016*, 5012341. https://doi.org/10.1155/2016/5012341. Epub 2016 Dec 8.

Akiyama, T., Nagamine, M., Davoodi, A., Ivanov, M., Carstens, M. I., & Carstens, E. (2017). Innocuous warming enhances peripheral serotonergic itch signaling and evokes enhanced responses in serotonin-responsive dorsal horn neurons in the mouse. *Journal of Neurophysiology, 117*(1), 251–259. https://doi.org/10.1152/jn.00703.2016. Epub 2016 Oct 26.

Anderson, K., Aito, S., Atkins, M., Biering-Sørensen, F., Charlifue, S., Curt, A., et al. (2008). Functional recovery measures for spinal cord injury: An evidence-based review for clinical practice and research. *The Journal of Spinal Cord Medicine, 31*(2), 133–144.

Bela, S., Wong, A. S., & Ho, G. W. (2010). Hydrolysis and ion exchange of titania nanoparticles towards large-scale titania and titanate nanobelts for gas sensing applications. *Journal of Physics D: Applied Physics, 43*(3) 035401.

Biggs, M. J., Richards, R. G., & Dalby, M. J. (2010). Nanotopographical modification: A regulator of cellular function through focal adhesions. *Nanomedicine: Nanotechnology, Biology and Medicine, 6*(5), 619–633.

Blight, A. (1988). Mechanical factors in experimental spinal cord injury. *The Journal of the American Paraplegia Society, 11*(2), 26–34. Review.

Bunge, M. B., & Wood, P. M. (2012). Realizing the maximum potential of Schwann cells to promote recovery from spinal cord injury. *Handbook of Clinical Neurology, 109*, 523–540. https://doi.org/10.1016/B978-0-444-52137-8.00032-2. Review.

Chedly, J., Soares, S., Montembault, A., von Boxberg, Y., Veron-Ravaille, M., Mouffle, C., Benassy, M. N., Taxi, J., David, L., & Nothias, F. (2017). Physical chitosan

microhydrogels as scaffolds for spinal cord injury restoration and axon regeneration. *Biomaterials*, *138*, 91–107.
Chen, X., & Mao, S. S. (2007). Titanium dioxide nanomaterials: Synthesis, properties, modifications, and applications. *Chemical Reviews*, *107*(7), 2891–2959.
Cheng, X. H., Zhang, L., & Fu, J. (2019). Somatosensory evoked potential changes and decompression timing for spinal cord function recovery and evoked potentials in rats with spinal cord injury. *Brain Research Bulletin*, *146*, 7–11. https://doi.org/10.1016/j.brainresbull.2018.12.003. Epub 2018 Dec 11.
Choi, W., Choi, K., Yang, G., Kim, J. C., & Yu, C. (2016). Improving piezoelectric performance of lead-free polymer composites with high aspect ratio BaTiO3 nanowires. *Polymer Testing*, *53*, 143–148.
Chu, L., Qin, Z., & Yang, J. (2015). Anatase TiO2 nanoparticles with exposed {001} facets for efficient dye-sensitized solar cells. *Scientific Reports*, *5*, 12143.
Côté, M. P., Amin, A. A., Tom, V. J., & Houle, J. D. (2011). Peripheral nerve grafts support regeneration after spinal cord injury. *Neurotherapeutics*, *8*(2), 294–303. https://doi.org/10.1007/s13311-011-0024-6. Review.
Cui, L. Y., Zhu, Y. C., Gao, S., Wang, J. M., Peng, B., Ni, J., et al. (2013). Ninety-day administration of dl-3-n-butylphthalide for acute ischemic stroke: A randomized, double-blind trial. *Chinese Medical Journal*, *126*(18), 3405–3410.
Dalby, M. J., McCloy, D., Robertson, M., Agheli, H., Sutherland, D., Affrossman, S., et al. (2006). Osteoprogenitor response to semi-ordered and random nanotopographies. *Biomaterials*, *27*(15), 2980–2987.
Dallacasa, V. (2012). Enhanced size-dependent piezoelectricity in nanostructured TiO2 films. *ISRN Materials Science*, *8*, 2012.
de la Torre, J. C. (1981). Spinal cord injury. Review of basic and applied research. *Spine (Phila Pa 1976)*, *6*(4), 315–335. Review.
Deecke, L., & Tator, C. H. (1973). Neurophysiological assessment of afferent and efferent conduction in the injured spinal cord of monkeys. *Journal of Neurosurgery*, *39*(1), 65–74. No abstract available.
Dietz, V., & Schwab, M. E. (2017). From the rodent spinal cord injury model to human application: Promises and challenges. *Journal of Neurotrauma*, *34*(9), 1826–1830. https://doi.org/10.1089/neu.2016.4513. Epub 2016 Dec 21. Review.
Dietz, V., Wirz, M., & Jensen, L. (1997). Locomotion in patients with spinal cord injuries. *Physical Therapy*, *77*(5), 508–516. Review.
Dimitrijevic, M. R., Danner, S. M., & Mayr, W. (2015). Neurocontrol of movement in humans with spinal cord injury. *Artificial Organs*, *39*(10), 823–833. https://doi.org/10.1111/aor.12614. Review.
Dimitrijevic, M. R., Faganel, J., & Young, R. R. (1981). Underlying mechanisms of the effects of spinal cord stimulation in motor disorders. A review of the discussion. *Applied Neurophysiology*, *44*(1–3), 133–140.
Dimitrijevic, M. R., Hsu, C. Y., & McKay, W. B. (1992). Neurophysiological assessment of spinal cord and head injury. *Journal of Neurotrauma*, *9*(Suppl 1), S293–S300. Review.
Dinan, B., Gallego-Perez, D., Lee, H., Hansford, D., & Akbar, S. A. (2013). Thermally grown TiO2 nanowires to improve cell growth and proliferation on titanium based materials. *Ceramics International*, *39*(5), 5949–5954.
Ditunno, J. F., Jr. (1997). Functional outcomes in spinal cord injury (SCI): Quality care versus cost containment. *The Journal of Spinal Cord Medicine*, *20*(1), 1–7. Review.
Ducker, T. B., Salcman, M., Lucas, J. T., Garrison, W. B., & Perot, P. L., Jr. (1978). Experimental spinal cord trauma, II: Blood flow, tissue oxygen, evoked potentials in both paretic and plegic monkeys. *Surgical Neurology*, *10*(1), 64–70.
Dupuis, A. C., Jodin, L., & Rouvière, E. (2006). Catalytic growth of TiO2 nanowires from a TiN thin film. *Applied Surface Science*, *253*(3), 1227–1235.

Fateh, R., Dillert, R., & Bahnemann, D. (2014). Self-Cleaning properties, mechanical stability, and adhesion strength of transparent photocatalytic TiO2–ZnO coatings on polycarbonate. *ACS Applied Materials & Interfaces*, *6*(4), 2270–2278.

Feng, L., Sharma, A., Niu, F., Huang, Y., Lafuente, J. V., Muresanu, D. F., et al. (2018). TiO_2-nanowired delivery of DL-3-n-butylphthalide (DL-NBP) attenuates blood-brain barrier disruption, brain edema formation, and neuronal damages following concussive head injury. *Molecular Neurobiology*, *55*(1), 350–358. https://doi.org/10.1007/s12035-017-0746-5.

Führmann, T., Tam, R. Y., Ballarin, B., Coles, B., Elliott Donaghue, I., van der Kooy, D., et al. (2016). Injectable hydrogel promotes early survival of induced pluripotent stem cell-derived oligodendrocytes and attenuates longterm teratoma formation in a spinal cord injury model. *Biomaterials*, *83*, 23–36. https://doi.org/10.1016/j.biomaterials.2015.12.032. Epub 2016 Jan 1.

Führmann, T., et al. (2017). Combinatorial therapies after spinal cord injury: How can biomaterials help? *Advanced Healthcare Materials*, *6*.

Gad, P., Choe, J., Nandra, M. S., Zhong, H., Roy, R. R., Tai, Y. C., et al. (2013). Development of a multi-electrode array for spinal cord epidural stimulation to facilitate stepping and standing after a complete spinal cord injury in adult rats. *Journal of Neuroengineering and Rehabilitation*, *10*, 2. https://doi.org/10.1186/1743-0003-10-2. Erratum in: J Neuroeng Rehabil. 2015;12:33.

Ganzer, P. D., Beringer, C. R., Shumsky, J. S., Nwaobasi, C., & Moxon, K. A. (2018). Serotonin receptor and dendritic plasticity in the spinal cord mediated by chronic serotonergic pharmacotherapy combined with exercise following complete SCI in the adult rat. *Experimental Neurology*, *304*, 132–142. https://doi.org/10.1016/j.expneurol.2018.03.006. Epub 2018 Mar 9.

Gardner, B. P., Theocleous, F., & Krishnan, K. R. (1988). Outcome following acute spinal cord injury: A review of 198 patients. *Paraplegia*, *26*(2), 94–98.

Gavaret, M., Pesenti, S., Diop-Sene, M. S., Choufani, E., Bollini, G., & Jouve, J. L. (2017). Intraoperative spinal cord monitoring: Lesional level diagnosis. *Orthopaedics & Traumatology, Surgery & Research*, *103*(1), 33–38. https://doi.org/10.1016/j.otsr.2016.09.021. Epub 2016 Dec 14.

Ghosh, M., & Pearse, D. D. (2015). The role of the serotonergic system in locomotor recovery after spinal cord injury. *Frontiers in Neural Circuits*, *8*, 151. https://doi.org/10.3389/fncir.2014.00151. eCollection 2014. Review.

Goriainov, V., Hulsart-Billstrom, G., Sjostrom, T., Dunlop, D. G., Su, B., & Oreffo, R. O. (2018). Harnessing nanotopography to enhance osseointegration of clinical orthopedic titanium implants—An in vitro and in vivo analysis. *Frontiers in Bioengineering and Biotechnology*, *6*, 44.

Grundy, B. L., & Friedman, W. (1987). Electrophysiological evaluation of the patient with acute spinal cord injury. *Critical Care Clinics*, *3*(3), 519–548. Review.

He, Z., Zhou, Y., Huang, Y., Wang, Q., Zheng, B., Zhang, H., et al. (2017). Dl-3-n-butylphthalide improves functional recovery in rats with spinal cord injury by inhibiting endoplasmic reticulum stress-induced apoptosis. *American Journal of Translational Research*, *9*(3), 1075–1087. eCollection 2017.

He, Z., Zhou, Y., Lin, L., Wang, Q., Khor, S., Mao, Y., et al. (2017). Dl-3-n-butylphthalide attenuates acute inflammatory activation in rats with spinal cord injury by inhibiting microglial TLR4/NF-κB signalling. *Journal of Cellular and Molecular Medicine*, *21*(11), 3010–3022. https://doi.org/10.1111/jcmm.13212. Epub 2017 Aug 25.

Hu, H. Z., Jeffery, N. D., & Granger, N. (2018). Somatosensory and motor evoked potentials in dogs with chronic severe thoracolumbar spinal cord injury. *Veterinary Journal*, *237*, 49–54. https://doi.org/10.1016/j.tvjl.2018.05.007. Epub 2018 May 25.

Hughes, J. T. (1984). Regeneration in the human spinal cord: A review of the response to injury of the various constituents of the human spinal cord. *Paraplegia*, *22*(3), 131–137. No abstract available.

Hurlbert, R. J., Tator, C. H., & Theriault, E. (1993). Dose-response study of the pathological effects of chronically applied direct current stimulation on the normal rat spinal cord. *Journal of Neurosurgery*, *79*(6), 905–916.

Ievins, A., & Moritz, C. T. (2017). Therapeutic stimulation for restoration of function after spinal cord injury. *Physiology (Bethesda, Md)*, *32*(5), 391–398. https://doi.org/10.1152/physiol.00010.2017.

Jia, J., Wei, C., Liang, J., Zhou, A., Zuo, X., Song, H., et al. (2016). The effects of DL-3-n-butylphthalide in patients with vascular cognitive impairment without dementia caused by subcortical ischemic small vessel disease: A multicentre, randomized, double-blind, placebo-controlled trial. *Alzheimer's & Dementia*, *12*(2), 89–99. https://doi.org/10.1016/j.jalz.2015.04.010. Epub 2015 Jun 15.

Kakulas, A. (1988). The applied neurobiology of human spinal cord injury: A review. *Paraplegia*, *26*(6), 371–379. Review.

Kalinina, N. I., Zaitsev, A. V., & Vesselkin, N. P. (2018). Presynaptic serotonin 5-$HT_{1B/D}$ receptor-mediated inhibition of glycinergic transmission to the frog spinal motoneurons. *Journal of Comparative Physiology A, Neuroethology, Sensory, Neural, and Behavioral Physiology*, *204*(3), 329–337. https://doi.org/10.1007/s00359-017-1244-y. Epub 2017 Dec 30.

Kasuga, T., Hiramatsu, M., Hoson, A., Sekino, T., & Niihara, K. (1998). Formation of titanium oxide nanotube. *Langmuir*, *14*(12), 3160–3163.

Kasuga, T., Hiramatsu, M., Hoson, A., Sekino, T., & Niihara, K. (1999). Titania nanotubes prepared by chemical processing. *Advanced Materials*, *11*(15), 1307–1311.

Kim, H. M., Miyaji, F., Kokubo, T., Nishiguchi, S., & Nakamura, T. (1999). Graded surface structure of bioactive titanium prepared by chemical treatment. *Journal of Biomedical Materials Research*, *45*(2), 100–107.

Kobrine, A. I., Evans, D. E., & Rizzoli, H. V. (1978). Correlation of spinal cord blood flow, sensory evoked response, and spinal cord function in subacute experimental spinal cord compression. *Advanced Neurology*, *20*, 389–394. No abstract available.

Kojima, Y., Yamamoto, T., Ogino, H., Okada, K., & Ono, K. (1979). Evoked spinal potentials as a monitor of spinal cord viability. *Spine (Phila Pa 1976)*, *4*(6), 471–477.

Krishna, V., Konakondla, S., Nicholas, J., Varma, A., Kindy, M., & Wen, X. (2013). Biomaterial-based interventions for neuronal regeneration and functional recovery in rodent model of spinal cord injury: A systematic review. *The Journal of Spinal Cord Medicine*, *36*(3), 174–190. https://doi.org/10.1179/2045772313Y.0000000095.

Lakshmi, B. B., Patrissi, C. J., & Martin, C. R. (1997). Sol–gel template synthesis of semiconductor oxide micro-and nanostructures. *Chemistry of Materials*, *9*(11), 2544–2550.

Lee, K. H., Kim, U. J., Park, Y. G., Won, R., Lee, H., & Lee, B. H. (2011). Optical imaging of somatosensory evoked potentials in the rat cerebral cortex after spinal cord injury. *Journal of Neurotrauma*, *28*(5), 797–807. https://doi.org/10.1089/neu.2010.1492. Epub 2011 Apr 12.

Lei, Y., Zhang, L. D., & Fan, J. C. (2001). Fabrication, characterization and Raman study of TiO2 nanowire arrays prepared by anodic oxidative hydrolysis of TiCl3. *Chemical Physics Letters*, *338*(4–6), 231–236.

Lei, Y., Zhang, L. D., Meng, G. W., Li, G. H., Zhang, X. Y., Liang, C. H., et al. (2001). Preparation and photoluminescence of highly ordered TiO2 nanowire arrays. *Applied Physics Letters*, *78*(8), 1125–1127.

Lin, Y., Wu, G. S., Yuan, X. Y., Xie, T., & Zhang, L. D. (2003). Fabrication and optical properties of TiO2 nanowire arrays made by sol–gel electrophoresis deposition into anodic alumina membranes. *Journal of Physics: Condensed Matter*, *15*(17), 2917.

Hong, L. T. A., et al. (2017). An injectable hydrogel enhances tissue repair after spinal cord injury by promoting extracellular matrix remodeling. *Nature Communications*, *8*, 533.

Ma, V. Y., Chan, L., & Carruthers, K. J. (2014). Incidence, prevalence, costs, and impact on disability of common conditions requiring rehabilitation in the United States: Stroke, spinal cord injury, traumatic brain injury, multiple sclerosis, osteoarthritis, rheumatoid arthritis, limb loss, and back pain. *Archives of Physical Medicine and Rehabilitation*, *95*(5), 986–995.e1. https://doi.org/10.1016/j.apmr.2013.10.032. Epub 2014 Jan 21.

Ma, R. C., & Dun, N. J. (1986). Excitation of lateral horn neurons of the neonatal rat spinal cord by 5-hydroxytryptamine. *Brain Research*, *389*(1–2), 89–98.

Marino, A. A., & Becker, R. O. (1970). Piezoelectric effect and growth control in bone. *Nature*, *228*(5270), 473.

Martin, C. R. (1994). Nanomaterials: A membrane-based synthetic approach. *Science*, *266*(5193), 1961–1966.

Martin, S. H., & Bloedel, J. R. (1973). Evaluation of experimental spinal cord injury using cortical evoked potentials. *Journal of Neurosurgery*, *39*(1), 75–81. No abstract available.

McKay, W. B., Stokic, D. S., & Dimitrijevic, M. R. (1997). Assessment of corticospinal function in spinal cord injury using transcranial motor cortex stimulation: A review. *Journal of Neurotrauma*, *14*(8), 539–548. Review.

Meinecke, F. W. (1985). Some thoughts about neurological recovery in spinal cord injuries: A philosophical review. *Paraplegia*, *23*(2), 78–81.

Mesbah, S., Angeli, C. A., Keynton, R. S., El-Baz, A., & Harkema, S. J. (2017). A novel approach for automatic visualization and activation detection of evoked potentials induced by epidural spinal cord stimulation in individuals with spinal cord injury. *PLoS One*, *12*(10), e0185582. https://doi.org/10.1371/journal.pone.0185582. eCollection 2017.

Miao, Z., Xu, D., Ouyang, J., Guo, G., Zhao, X., & Tang, Y. (2002). Electrochemically induced sol–gel preparation of single-crystalline TiO2 nanowires. *Nano Letters*, *2*(7), 717–720.

Minassian, K., & Hofstoetter, U. S. (2016). Spinal cord stimulation and augmentative control strategies for leg movement after spinal paralysis in humans. *CNS Neuroscience & Therapeutics*, *22*(4), 262–270. https://doi.org/10.1111/cns.12530. Epub 2016 Feb 18. Review.

Morgado, E., de Abreu, M. A., Moure, G. T., Marinkovic, B. A., Jardim, P. M., & Araujo, A. S. (2007). Characterization of nanostructured titanates obtained by alkali treatment of TiO2-anatases with distinct crystal sizes. *Chemistry of Materials*, *19*(4), 665–676.

Nash, C. L., Jr., & Brown, R. H. (1989). Spinal cord monitoring. *The Journal of Bone and Joint Surgery American Volume*, *71*(4), 627–630. Review. No abstract available.

New, P. W. (2005). Functional outcomes and disability after nontraumatic spinal cord injury rehabilitation: Results from a retrospective study. *Archives of Physical Medicine and Rehabilitation*, *86*(2), 250–261.

Niu, F., Sharma, A., Feng, L., Ozkizilcik, A., Muresanu, D. F., Lafuente, J. V., et al. (2019). Nanowired delivery of DL-3-n-butylphthalide induces superior neuroprotection in concussive head injury. *Progress in Brain Research*, *245*, 89–118. https://doi.org/10.1016/bs.pbr.2019.03.008. Epub 2019 Apr 2.

Nógrádi, A., Pajer, K., & Márton, G. (2011). The role of embryonic motoneuron transplants to restore the lost motor function of the injured spinal cord. *Annals of Anatomy*, *193*(4), 362–370. https://doi.org/10.1016/j.aanat.2011.04.001. Epub 2011 Apr 30. Review.

Nor, A. M., Achoi, M. F., Mamat, M. H., Zabidi, M. M., Abdullah, S., & Mahmood, M. R. (2012). Synthesis of TiO2 nanowires via hydrothermal method. *Japanese Journal of Applied Physics*, *51*(6S), 06FG08.

Nuwer, M. R. (1988). Use of somatosensory evoked potentials for intraoperative monitoring of cerebral and spinal cord function. *Neurologic Clinics*, *6*(4), 881–897. Review.
Nyberg, F., Sharma, H. S., & Wisenfeld-Hallin, Z. (1995). Neuropeptides in the spinal cord, progress in brain research. In (Vol. 104)Amsterdam: Elsevier, pp. 1–430.
Othman, M. A., Amat, N. F., Ahmad, B. H., & Rajan, J. (2014). Electrical conductivity characteristic of TiO2 nanowires from hydrothermal method. In *Journal of Physics: Conference Series* (p. 012027). , Vol. 495 (p. 012027). IOP Publishing. No. 1.
Pakulska, M. M., Ballios, B. G., & Shoichet, M. S. (2012). Injectable hydrogels for central nervous system therapy. *Biomedical Materials*, 7(2), 024101. https://doi.org/10.1088/1748-6041/7/2/024101. Epub 2012 Mar 29.
Perot, P. L., Jr. (1973). Chapter 28. The clinical use of somatosensory evoked potentials in spinal cord injury. *Clinical Neurosurgery*, *20*, 367–381. No abstract available.
Puchala, E., & Windle, W. F. (1977). The possibility of structural and functional restitution after spinal cord injury. A review. *Experimental Neurology*, *55*(1), 1–42. No abstract available.
Roduner, E. (2006). Size matters: Why nanomaterials are different. *Chemical Society Reviews*, *35*(7), 583–592.
Rowed, D. W., McLean, J. A., & Tator, C. H. (1978). Somatosensory evoked potentials in acute spinal cord injury: Prognostic value. *Surgical Neurology*, *9*(3), 203–210.
Sadowsky, C. L. (2001). Electrical stimulation in spinal cord injury. *NeuroRehabilitation*, *16*(3), 165–169.
Schnell, L., & Schwab, M. E. (1990). Axonal regeneration in the rat spinal cord produced by an antibody against myelin-associated neurite growth inhibitors. *Nature*, *343*(6255), 269–272.
Schnell, L., & Schwab, M. E. (1993). Sprouting and regeneration of lesioned corticospinal tract fibres in the adult rat spinal cord. *The European Journal of Neuroscience*, *5*(9), 1156–1171.
Schramm, J. (1985). Spinal cord monitoring: Current status and new developments. *Central Nervous System Trauma*, *2*(3), 207–227. Review.
Schwab, M. E. (1992). Regeneration of lesioned CNS axons by neutralization of neurite growth inhibitors: A short review. *Journal of Neurotrauma*, *9*(Suppl 1), S219–S221. Review.
Schwab, M. E. (1993). Experimental aspects of spinal cord regeneration. *Current Opinion in Neurology and Neurosurgery*, *6*(4), 549–553. Review.
Sharma, H. S. (2004). Pathophysiology of the blood-spinal cord barrier in traumatic injury. In H. S. Sharma & J. Westman (Eds.), *The blood-spinal cord and brain barriers in health and disease* (pp. 437–518). San Diego: Elsevier Academic Press.
Sharma, H. S. (2007). A select combination of neurotrophins enhances neuroprotection and functional recovery following spinal cord injury. *Annals of the New York Academy of Sciences*, *1122*, 95–111.
Sharma, H. S. (2008). New perspectives for the treatment options in spinal cord injury. *Expert Opinion on Pharmacotherapy*, *9*(16), 2773–2800. https://doi.org/10.1517/14656566.9.16.2773. Review.
Sharma, H. S. (2010). Selected combination of neurotrophins potentiate neuroprotection and functional recovery following spinal cord injury in the rat. *Acta Neurochirurgica Supplement*, *106*, 295–300. https://doi.org/10.1007/978-3-211-98811-4_55.
Sharma, H. S., Ali, S. F., Dong, W., Tian, Z. R., Patnaik, R., Patnaik, S., et al. (2007). Drug delivery to the spinal cord tagged with nanowire enhances neuroprotective efficacy and functional recovery following trauma to the rat spinal cord. *Annals of the New York Academy of Sciences*, *1122*, 197–218.
Sharma, H. S., Ali, S., Tian, Z. R., Patnaik, R., Patnaik, S., Lek, P., et al. (2009). Nano-drug delivery and neuroprotection in spinal cord injury. *Journal of Nanoscience and Nanotechnology*, *9*(8), 5014–5037.

Sharma, H. S., Nyberg, F., & Olsson, Y. (1992). Dynorphin A content in the rat brain and spinal cord after a localized trauma to the spinal cord and its modification with p-chlorophenylalanine. An experimental study using radioimmunoassay technique. *Neuroscience Research*, *14*(3), 195–203.

Sharma, H. S., Nyberg, F., Westman, J., Alm, P., Gordh, T., & Lindholm, D. (1998). Brain derived neurotrophic factor and insulin like growth factor-1 attenuate upregulation of nitric oxide synthase and cell injury following trauma to the spinal cord. An immunohistochemical study in the rat. *Amino Acids*, *14*(1–3), 121–129.

Sharma, H. S., & Olsson, Y. (1990). Edema formation and cellular alterations following spinal cord injury in the rat and their modification with p-chlorophenylalanine. *Acta Neuropathologica*, *79*(6), 604–610.

Sharma, H. S., Olsson, Y., & Cervós-Navarro, J. (1993). p-Chlorophenylalanine, a serotonin synthesis inhibitor, reduces the response of glial fibrillary acidic protein induced by trauma to the spinal cord. An immunohistochemical investigation in the rat. *Acta Neuropathologica*, *86*(5), 422–427.

Sharma, H. S., Olsson, Y., & Dey, P. K. (1990). Early accumulation of serotonin in rat spinal cord subjected to traumatic injury. Relation to edema and blood flow changes. *Neuroscience*, *36*(3), 725–730.

Sharma, H. S., Olsson, Y., Nyberg, F., & Dey, P. K. (1993). Prostaglandins modulate alterations of microvascular permeability, blood flow, edema and serotonin levels following spinal cord injury: An experimental study in the rat. *Neuroscience*, *57*(2), 443–449.

Sharma, H. S., Olsson, Y., & Westman, J. (1995). A serotonin synthesis inhibitor, p-chlorophenylalanine reduces the heat shock protein response following trauma to the spinal cord: An immunohistochemical and ultrastructural study in the rat. *Neuroscience Research*, *21*(3), 241–249.

Sharma, H. S., & Westman, J. (2004). *The blood-spinal cord and brain barriers in health and disease*. (pp. 1–617). San Diego: Academic Press. Release date: Nov. 9 2003.

Sharma, H. S., & Winkler, T. (2002). Assessment of spinal cord pathology following trauma using early changes in the spinal cord evoked potentials: A pharmacological and morphological study in the rat. *Muscle & Nerve. Supplement*, *11*, S83–S91.

Sharma, H. S., Winkler, T., Stålberg, E., Olsson, Y., & Dey, P. K. (1991). Evaluation of traumatic spinal cord edema using evoked potentials recorded from the spinal epidural space. An experimental study in the rat. *Journal of the Neurological Sciences*, *102*(2), 150–162.

Stålberg, E., Sharma, H. S., & Olsson, Y. (1998). Spinal cord monitoring. In *Basic principles, regeneration, pathophysiology and clinical aspects* (pp. 1–527). Wien New York: Springer.

Tamaki, T., Tsuji, H., Inoue, S., & Kobayashi, H. (1981). The prevention of iatrogenic spinal cord injury utilizing the evoked spinal cord potential. *International Orthopaedics*, *4*(4), 313–317.

Tator, C. H. (1972). Acute spinal cord injury: A review of recent studies of treatment and pathophysiology. *Canadian Medical Association Journal*, *107*(2), 143–145 passim. Review. No abstract available.

Tator, C. H., & Fehlings, M. G. (1991). Review of the secondary injury theory of acute spinal cord trauma with emphasis on vascular mechanisms. *Journal of Neurosurgery*, *75*(1), 15–26. Review.

Theisen, C. C., Sachdeva, R., Austin, S., Kulich, D., Kranz, V., & Houle, J. D. (2017). Exercise and peripheral nerve grafts as a strategy to promote regeneration after acute or chronic spinal cord injury. *Journal of Neurotrauma*, *34*(10), 1909–1914. https://doi.org/10.1089/neu.2016.4640. Epub 2017 Apr 26. Review.

Thirumala, P., Zhou, J., Krishnan, R., Manem, N., Umredkar, S., Hamilton, D. K., et al. (2016). Diagnostic accuracy of evoked potentials for functional impairment after contusive spinal cord injury in adult rats. *Journal of Clinical Neuroscience*, *25*, 122–126. https://doi.org/10.1016/j.jocn.2015.10.010. Epub 2015 Dec 8.

Thomaz, S. R., Cipriano, G., Jr., Formiga, M. F., Fachin-Martins, E., Cipriano, G. F. B., Martins, W. R., et al. (2019). Effect of electrical stimulation on muscle atrophy and spasticity in patients with spinal cord injury—A systematic review with meta-analysis. *Spinal Cord*, *57*(4), 258–266. https://doi.org/10.1038/s41393-019-0250-z. Epub 2019 Jan 29. Review.

Tidoni, E., Tieri, G., & Aglioti, S. M. (2015). Re-establishing the disrupted sensorimotor loop in deafferented and deefferented people: The case of spinal cord injuries. *Neuropsychologia*, *79*(Pt B), 301–309. https://doi.org/10.1016/j.neuropsychologia.2015.06.029, Epub 2015 Jun 24.

Tsai, C. C., & Teng, H. (2006). Structural features of nanotubes synthesized from NaOH treatment on TiO2 with different post-treatments. *Chemistry of Materials*, *18*(2), 367–373.

Valden, M., Lai, X., & Goodman, D. W. (1998). Onset of catalytic activity of gold clusters on titania with the appearance of nonmetallic properties. *Science*, *281*(5383), 1647–1650.

van Hove, R. P., Sierevelt, I. N., van Royen, B. J., & Nolte, P. A. (2015). Titanium-nitride coating of orthopaedic implants: A review of the literature. *BioMed Research International*, *2015*, 485975.

Veneruso, V., Rossi, F., Villella, A., Bena, A., Forloni, G., & Veglianese, P. (2019). Stem cell paracrine effect and delivery strategies for spinal cord injury regeneration. *Journal of Controlled Release*, *300*, 141–153. https://doi.org/10.1016/j.jconrel.2019.02.038. Epub 2019 Mar 6. Review.

Wang, Y., Bi, Y., Xia, Z., Shi, W., Li, B., Li, B., et al. (2018). Butylphthalide ameliorates experimental autoimmune encephalomyelitis by suppressing PGAM5-induced necroptosis and inflammation in microglia. *Biochemical and Biophysical Research Communications*, *497*(1), 80–86. https://doi.org/10.1016/j.bbrc.2018.02.024. Epub 2018 Feb 3.

Whelan, A., Halpine, M., Christie, S. D., & McVeigh, S. A. (2018). Systematic review of melatonin levels in individuals with complete cervical spinal cord injury. *The Journal of Spinal Cord Medicine*, *22*, 1–14. https://doi.org/10.1080/10790268.2018.1505312. Epub ahead of print.

Wilson, R. D., Bryden, A. M., Kilgore, K. L., Makowski, N., Bourbeau, D., Kowalski, K. E., et al. (2019). Neuromodulation for functional electrical stimulation. *Physical Medicine and Rehabilitation Clinics of North America*, *30*(2), 301–318. https://doi.org/10.1016/j.pmr.2018.12.011. Review.

Winkler, T., Sharma, H. S., Gordh, T., Badgaiyan, R. D., Stålberg, E., & Westman, J. (2002). Topical application of dynorphin A (1-17) antiserum attenuates trauma induced alterations in spinal cord evoked potentials, microvascular permeability disturbances, edema formation and cell injury: An experimental study in the rat using electrophysiological and morphological approaches. *Amino Acids*, *23*(1–3), 273–281.

Winkler, T., Sharma, H. S., Stålberg, E., & Badgaiyan, R. D. (2000). Neurotrophic factors attenuate alterations in spinal cord evoked potentials and edema formation following trauma to the rat spinal cord. *Acta Neurochirurgica. Supplement*, *76*, 291–296.

Winkler, T., Sharma, H. S., Stålberg, E., Badgaiyan, R. D., Alm, P., & Westman, J. (1998). Spinal cord evoked potentials and edema in the pathophysiology of rat spinal cord injury. Involvement of nitric oxide. *Amino Acids*, *14*(1–3), 131–139.

Winkler, T., Sharma, H. S., Stålberg, E., Badgaiyan, R. D., Gordh, T., & Westman, J. (2003). An L-type calcium channel blocker, nimodipine influences trauma induced spinal cord conduction and axonal injury in the rat. *Acta Neurochirurgica. Supplement*, *86*, 425–432.

Winkler, T., Sharma, H. S., Stålberg, E., Badgaiyan, R. D., Westman, J., & Nyberg, F. (2000). Growth hormone attenuates alterations in spinal cord evoked potentials and cell injury following trauma to the rat spinal cord. An experimental study using topical application of rat growth hormone. *Amino Acids*, *19*(1), 363–371.

Winkler, T., Sharma, H. S., Stålberg, E., & Olsson, Y. (1993). Indomethacin, an inhibitor of prostaglandin synthesis attenuates alteration in spinal cord evoked potentials and edema formation after trauma to the spinal cord: An experimental study in the rat. *Neuroscience*, *52*(4), 1057–1067.

Winkler, T., Sharma, H. S., Stålberg, E., Olsson, Y., & Nyberg, F. (1994a). Opioid receptors influence spinal cord electrical activity and edema formation following spinal cord injury: Experimental observations using naloxone in the rat. *Neuroscience Research*, *21*(1), 91–101.

Winkler, T., Sharma, H. S., Stålberg, E., Olsson, Y., & Nyberg, F. (1994b). Naloxone reduces alterations in evoked potentials and edema in trauma to the rat spinal cord. *Acta Neurochirurgica. Supplementum (Wien)*, *60*, 511–515.

Winkler, T., Sharma, H. S., Stålberg, E., Olsson, Y., & Nyberg, F. (1995). Role of histamine in spinal cord evoked potentials and edema following spinal cord injury: Experimental observations in the rat. *Inflammation Research*, *44*(Suppl. 1), S44–S45. No abstract available.

Winkler, T., Sharma, H. S., Stålberg, E., & Westman, J. (1997). Benzodiazepine receptors influence spinal cord evoked potentials and edema following trauma to the rat spinal cord. *Acta Neurochirurgica. Supplement*, *70*, 216–219.

Winkler, T., Sharma, H. S., Stålberg, E., Westman, Stålberg, E., Sharma, H. S., et al. (1998). Spinal cord bioelectrical activity, edema and cell injury following a focal trauma to the spinal cord. An experimental study using pharmacological and morphological approach. In *Spinal cord monitoring. Basic principles, regeneration, pathophysiology and clinical aspects* (pp. 281–348). Wien New York: Springer.

Wood-Dauphinée, S., Exner, G., Bostanci, B., Exner, G., Glass, C., Jochheim, K. A., et al. (2002). Quality of life in patients with spinal cord injury—Basic issues, assessment, and recommendations. *Restorative Neurology and Neuroscience*, *20*(3–4), 135–149. Review.

Woolsey, R. M. (1988). Modern concepts of therapy and management of spinal cord injuries. *Critical Reviews in Neurobiology*, *4*(2), 137–156. Review.

Wu, Y., Long, M., Cai, W., Dai, S., Chen, C., Wu, D., et al. (2009). Preparation of photocatalytic anatase nanowire films by in situ oxidation of titanium plate. *Nanotechnology*, *20*(18), 185703.

Wu, F., Yu, Y., Yang, H., German, L. N., Li, Z., Chen, J., et al. (2017). Simultaneous enhancement of charge separation and hole transportation in a TiO2–SrTiO3 core–shell nanowire photoelectrochemical system. *Advanced Materials*, *29*(28), 1701432.

Yang, H., He, B. R., & Hao, D. J. (2015). Biological roles of olfactory ensheathing cells in facilitating neural regeneration: A systematic review. *Molecular Neurobiology*, *51*(1), 168–179. https://doi.org/10.1007/s12035-014-8664-2. Epub 2014 Mar 11. Review.

Yang, H. G., Sun, C. H., Qiao, S. Z., Zou, J., Liu, G., Smith, S. C., et al. (2008). Anatase TiO 2 single crystals with a large percentage of reactive facets. *Nature*, *453*(7195), 638.

Yu, C., Xia, K., Gong, Z., Ying, L., Shu, J., Zhang, F., et al. (2019). The application of neural stem/progenitor cells for regenerative therapy of spinal cord injury. *Current Stem Cell Research & Therapy*. https://doi.org/10.2174/1574888X14666190329095638. Epub ahead of print.

Yuan, Z. Y., & Su, B. L. (2004). Titanium oxide nanotubes, nanofibers and nanowires. *Colloids and Surfaces A: Physicochemical and Engineering Aspects*, *241*(1–3), 173–183.

Zheng, B., Zhou, Y., Zhang, H., Yang, G., Hong, Z., Han, D., et al. (2017). Dl-3-n-butylphthalide prevents the disruption of blood-spinal cord barrier via inhibiting endoplasmic reticulum stress following spinal cord injury. *International Journal of Biological Sciences*, *13*(12), 1520–1531. https://doi.org/10.7150/ijbs.21107. eCollection 2017.

CHAPTER FOUR

Leakage of the blood-brain barrier followed by vasogenic edema as the ultimate cause of death induced by acute methamphetamine overdose

Eugene A. Kiyatkin[a,*], Hari Shanker Sharma[b]
[a]Behavioral Neuroscience Branch, National Institute on Drug Abuse—Intramural Research Program, National Institutes of Health, DHHS, Baltimore, MD, United States
[b]International Experimental Central Nervous System Injury & Repair (IECNSIR), Department of Surgical Sciences, Anesthesiology & Intensive Care Medicine, Uppsala University Hospital, Uppsala University, Uppsala, Sweden
*Corresponding author: e-mail address: ekiyatki@intra.nida.nih.gov

Contents

Abstract

Methamphetamine (METH) is a potent CNS stimulant that is widely used as a recreational drug. Due to its ability to increase bodily heat production and diminish heat loss due to peripheral vasoconstriction, METH is able to increase brain and body temperature. The hyperthermic effects of METH are potentiated when the drug is used under conditions of psycho-physiological activation and in warm ambient temperatures. In this short review, we present and discuss our data on the effects of METH on brain temperature and a number of neural parameters that characterize permeability of the blood-brain barrier (albumin immunoreactivity), glial activity (GFAP immunoreactivity), brain tissue water content, and structural abnormalities of brain cells. We demonstrate that the extent of these neural alterations strongly depends on METH-induced brain temperature elevation and they all dramatically increase following exposure to METH in warm (29 °C) vs. standard (23 °C) ambient temperatures. Based on these data we

International Review of Neurobiology, Volume 146
ISSN 0074-7742
https://doi.org/10.1016/bs.irn.2019.06.010

consider possible pathophysiological mechanisms underlying acute METH toxicity, suggesting the critical role of drug-induced brain hyperthermia, temperature-dependent leakage of the blood-brain barrier (BBB), and the development of vasogenic edema that could finally result in decompensation of vital functions and death.

Methamphetamine (METH) is a potent CNS stimulant that is widely used as a recreational drug. The current spike in the use of opioid drugs and the rise in lethality associated with these drugs shifted public attention away from METH and other related psychostimulants such as MDMA, which are still quite popular all over the world, especially with young users. While social and economic problems associated with drug abuse are well known, METH intake could result in serious health complications, including death following acute overdose. The typical symptoms of METH overdose in humans include arrhythmia, hyperventilation, hyperactivity, sweating, fever, muscular pain, and biphasic changes in arterial blood pressure (Prakash et al., 2017; Rusyniak & Sprague, 2005). Most of these symptoms reflect strong activation of the central nervous system (CNS) with subsequent hyperactivity of the sympathetic nervous system, which could be responsible for multiple autonomic effects of this drug. Among these effects, an extreme rise in body temperature (hyperthermia) is a prominent one, which could be responsible, at least in part, for the development of other pathological symptoms, including death.

It is generally believed that coma and lethality during METH overdose result primarily from heart attack or stroke. In this work, we aim to challenge this view and present several sets of data which suggest that brain hyperthermia, with subsequent leakage of the blood-brain barrier (BBB) and development of vasogenic edema, is the primary pathophysiological mechanism determining death during acute METH overdose. Although our data were obtained in rats, which have important differences from humans in temperature regulation (Gordon, 1990, 2007), we believe that certain findings of our studies could be applicable for understanding the acute toxic effects of METH in human drug users.

We start our discussion with a short introduction that defines brain temperature as an important homeostatic parameter, which simultaneously depends on metabolic neural activity and affects all manifestations of neural activity and neural function. While traditionally viewed as a strictly regulated parameter, brain temperature fluctuates within relatively wide limits under physiological conditions and is affected by different psychoactive drugs. Second, we will consider the brain temperature effects of METH and show how these effects depend on activity state and environmental conditions. Third, we will show that acute METH intoxication results in dramatic changes in BBB permeability, activity of glial cells, and brain water homeostasis; these changes are tightly interrelated and their extent depends on drug-induced temperature elevation. To conclude, we discuss functional implications of our studies in terms of understanding the acute toxic effects of METH and related substances. The primary data regarding changes in brain parameters during METH intoxication were the result of a collaboration between the National Institute on Drug Abuse-Intramural Research Program, USA and Uppsala University, Sweden.

1. Brain temperature: Basic mechanisms of brain temperature homeostasis

While it is traditionally believed that brain temperature in healthy homeothermic organisms is stable and close to 37 °C, relatively large fluctuations in brain temperature have been found to occur during different types of natural motivated behavior and following exposure to various environmental challenges (Abrams & Hammel, 1964; Baker, Frye, & Millet, 1973; Delgado & Hanai, 1966; Fuller & Baker, 1983; Hayward & Baker, 1968; Kovalson, 1972; McElligott & Melzack, 1967; Serota & Gerard, 1938). By using miniature thermocouple sensors chronically implanted in different brain structures, we showed that hypothalamic temperature in awake, freely moving male rats can fall to ~35 °C during deep sleep (Smirnov & Kiyatkin, 2008) and can phasically peak at ~39.5 °C at the time of ejaculation during copulatory behavior (Kiyatkin & Mitchum, 2003). These lower and upper limits obviously define the range of "normal" brain temperature fluctuations (~4 °C) within the entire physiological continuum.

Changes in brain temperature depend on two opposing forces. While metabolism-related intra-brain heat production tends to increase brain temperature, heat dissipation by cerebral blood flow to the rest of the body and then to the external environment tends to decrease it. Brain activity is highly energy consuming (Schmidt-Nielsen, 1997; Siesjo, 1978), and heat continuously generated by brain tissue is removed by cerebral circulation due to a temperature differential between arterial blood inflowing to the brain and brain tissue (Hayward & Baker, 1968; Kiyatkin, Brown, & Wise, 2002).

While it seems mechanistic, the cooling of an internal combustion engine appears to be a good analogy to brain temperature exchange. Similar to circulating coolant that continuously removes heat from a working engine, cool, oxygenated arterial blood removes heat from the brain via heat exchange and the now warmed venous blood returns to the heart to be cooled and oxygenated again in the lungs. Such an arrangement determines the critical role of cerebral blood flow in brain temperature homeostasis and the essential interdependence of temperature in the brain and the rest of the body. While physiological increases in brain temperature result from increased intra-brain heat production and diminished heat loss due to peripheral vasoconstriction, brain temperature can also rise when heat generated in brain tissue during "normal" metabolic activity cannot be properly dissipated to the body and then to the external environment. In this case, the extent of temperature increase is determined by the efficiency of heat outflow from the brain and the body, and this increase can exceed the physiological range (>39–40 °C).

While brain temperature is at its lowest values during natural sleep and is increased to a different extent following exposure to arousing stimuli and during motivated behaviors, brain temperature can also be decreased below its physiological range by drugs that have direct inhibiting actions on the CNS (i.e., general anesthetics, alcohol, benzodiazepines, natural and synthetic cannabinoids; see Kiyatkin, 2018 for review). As shown previously (Kiyatkin & Brown, 2005), general anesthesia induced by sodium pentobarbital induces robust decreases in brain temperature that occur due to coordinated changes in two factors: a decrease in intra-brain heat production and increased heat loss resulting from sustained skin vasodilation. Since most psychoactive drugs affect metabolism as well as the state of peripheral and cerebral blood vessels, they should affect brain temperature, and brain temperature responses should depend on the organism's ongoing state and environmental conditions. A drug at a certain dose could induce minimal temperature effects under one set of environmental conditions, in which adaptive mechanisms of heat loss are fully effective. However, the same drug at the same dose can induce pathological hyperthermia when used in different environmental conditions in which heat dissipation mechanisms are significantly impaired. Since peripheral vasodilatation and perspiration are powerful means for heat loss in humans, drug-induced impairment of these adaptive mechanisms could be a very important determinant of drug-induced increases in brain and body temperatures.

2. Brain hyperthermia induced by METH: State-dependency and environmental modulation

It is well known that METH induces dose-dependent body hyperthermia, which can reach clearly pathological levels following high dose exposure (Alberts & Sonsalla, 1995; Miller & O'Callaghan, 2003; Sandoval, Hanson, & Fleckenstein, 2000; Seiden & Sabol, 1996). It has been also shown that METH-induced hyperthermia is greatly potentiated in warm environmental temperatures that diminish natural heat dissipation. This phenomenon could be important for human conditions because METH is often used by young adults under conditions of physical and emotional activation, often in a warm and humid environment. While METH-induced metabolic activation and enhanced heat production in the brain and body is an essential factor determining hyperthermia, METH also induces strong peripheral vasoconstriction (Lynch & House, 1992), thus diminishing heat dissipation from body surfaces and enhancing heat accumulation in the brain.

To assess how METH affects brain temperature homeostasis and evaluate how the brain temperature effects of this drug are modulated by activity state and environmental conditions, we examined temperature changes in the nucleus accumbens (NAc), hippocampus and temporal muscle induced in male rats by METH (1–9 mg/kg, subcutaneously) in quiet resting conditions at a normal laboratory temperature (23 °C), during social interaction with a female, and at moderately warm ambient temperatures (29 °C) (Brown, Wise, & Kiyatkin, 2003). While most experiments in rats are typically conducted at standard laboratory temperatures (22–23 °C), 29 °C is not considered a truly hot temperature and it is within the range of normothermy or comfortable temperature in rats (Romanovsky, Ivanov, & Shimansky, 2002), where basal metabolism is maintained at its lowest levels. While we initially assessed the temperature effects of METH at a wide range of doses and showed that drug-induced temperature elevations are dose-dependent, for most subsequent studies METH was administered subcutaneously at a 9 mg/kg dose. Although this drug dose exceeds the doses typically used by humans, this dose corresponds to only 1/6 of the LD_{50} in rats (49 mg/kg for ip injection; Davis, Hatoum, & Walters, 1987; Yamamoto, 1963), and METH at this dose does not induce lethality under normal environmental conditions.

When tested at 29 °C, NAc temperature increased rapidly in all animals, rising to clearly pathological values (>41–42 °C) and resulting in death of 5 out of 6 animals within 6 h post-injection. Lethality resulting from METH at doses much lower than the LD_{50} indicates that standard procedures for determining acute toxicity can be misleading, and that much smaller drug doses used under specific environmental conditions can induce life-threatening health complications, including death (Fig. 1).

A powerful modulation of the temperature effects of METH by environmental conditions seen in rats may help to explain the exceptionally strong, sometimes fatal, responses of some individuals to amphetamine-like substances including METH and MDMA that occur under rave conditions. However, humans have much more efficient mechanisms of heat loss from body surfaces than do rats (Gordon, 1990, 2007), thus making them more resistant to high environmental temperatures and thermogenic effects of psychomotor stimulants. In contrast to rats, humans have a well-developed ability to sweat and have a very high dynamic range of blood flow rates in the skin, thus allowing them to lose more metabolic heat than can be maximally produced in the body (Rowell, 1983). These differences in the effector mechanisms of heat loss could explain weaker temperature increases induced by psychostimulants in monkeys (Banks et al., 2007; Taffe et al., 2006; Von Huben et al., 2007) and humans (Freedman, Johnson, & Tancer, 2005). Despite their high efficiency, the compensatory mechanisms of heat loss in humans could be greatly impaired under specific conditions, resulting in progressive heat accumulation in the organism. A simple bicycle exercise that induces a ~1 °C brain temperature elevation under normal conditions produces much stronger hyperthermia (39.0–39.5 °C) when the exercise is conducted in special water-impermeable clothing that prevents heat dissipation to the external environment (Nybo, Secher, & Nielson, 2002). Therefore, pathological brain hyperthermia induced by overdose of psychomotor stimulants under rave conditions results not only from excessive heat production due to drug-induced and psycho-physiological activation, but also from the impaired ability to dissipate metabolic heat due to powerful drug-induced peripheral vasoconstriction.

3. Changes in BBB permeability and brain water homeostasis during METH intoxication: Role of brain temperature

Considering the issue of neurotoxicity, it is usually assumed that METH and related drugs have direct toxic effects on neural cells, with

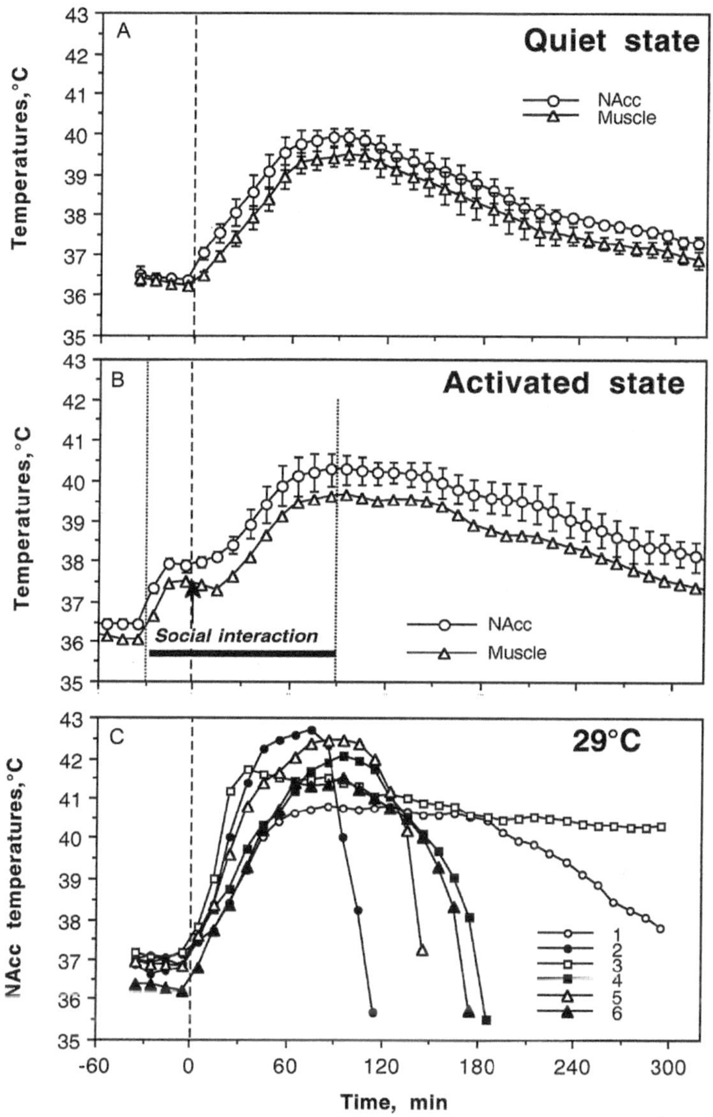

Fig. 1 Mean (±SEM) changes in temperature recorded from the nucleus accumbens (NAc) and temporal muscle following sc administration of METH (9 mg/kg) in freely-moving rats in quiet resting conditions (A), during social interaction with another rat (B) and in warm ambient temperatures (C). Vertical hatched lines show the moment of drug injection. Additional vertical dotted lines in graph B show the time interval of social interaction. *Original data were published in Brown, P.L., Wise, R.A., Kiyatkin, E.A. (2003). Brain hyperthermia is induced by methamphetamine and exacerbated by social interaction.* Journal of Neuroscience 23, *3924–3929.*

relative selectivity towards specific cell groups, brain structures, and cellular organelles. In particular, it was shown that METH has damaging effects on fine dopamine axonal terminals in the striatum (Ricaurte, Schuster, & Seiden, 1980; Riddle, Fleckenstein, & Hanson, 2006; Woolverton, Ricaurte, Forno, & Seiden, 1989) and this damage results in multiple health complications associated with diminished dopamine transmission. Alterations in the activity and responsiveness of dopamine and other monoamine systems could be important factors in psycho-emotional and psychiatric disorders including acute METH psychosis and severe depression following long-term METH use (Kalant, 2001). However, METH and other psychomotor stimulant drugs also induce strong metabolic activation and robust hyperthermia. Enhanced metabolism is tightly related to oxidative stress, which is caused by an accumulation of reactive oxygen species and an inability of an organism to detoxify them and repair the resulting damage. Disturbances in this normal redox state can cause toxic effects on brain cells through the production of peroxides and free radicals that damage all components of the cell including proteins, lipids, and DNA. Oxidative stress as a consequence of sustained metabolic activation is often viewed as a primary factor of METH-induced neurotoxicity (Cadet, Krasnova, Jayanthi, & Lyles, 2007; De Vito & Wagner, 1989; Stephans & Yamamoto, 1994). Since brain cells are exceptionally temperature-sensitive, with structural abnormalities appearing at ~40 °C, i.e., only three degrees above a normal baseline (Chen et al., 2003; Iwagami, 1996; Lepock, 2003; Oifa & Kleshchnov, 1985; Sharma & Hoopes, 2003; Yamamoto & Zhu, 1998), high temperature per se could be an important contributor to acute neurotoxicity. Due to the temperature dependence of most physico-chemical processes governing neural activity, hyperthermia could also be a powerful factor that enhances the toxic effects of METH and its metabolites on brain cells.

In addition to the direct effects of high temperatures on brain cells and the potentiation of toxic effects of drug metabolites, brain hyperthermia appears to alter permeability of the BBB—an important barrier that maintains stability of the brain environment and protects neural cells from potentially dangerous ionic and chemical perturbations occurring in the body (Rapoport, 1976; Zlokovic, 2008). Leakage of the BBB has been previously documented during environmental warming (Cervos-Navarro, Sharma, Westman, & Bongcum-Rudloff, 1998; Sharma, Zimmer, Westman, & Cervos-Navarro, 1992), intense physical exercise (Watson, Shirreffs, & Maughan, 2005), various types of stress (Esposito et al., 2001; Ovadia, Abramsky, Feldman, & Weidenfeld, 2001; Sharma & Dey, 1986), and morphine withdrawal (Sharma & Ali, 2006). In our studies, we examined how acute METH intoxication (9 mg/kg) affects BBB permeability, the state of glial cells, brain ion and water homeostasis, and structural abnormalities of different types of brain cells (Kiyatkin, Brown, & Sharma, 2007; Sharma & Kiyatkin, 2009).

For these experiments, rats were surgically implanted with a chronic temperature sensor in the NAc. After several days of habituation to the recording environment, rats were subcutaneously injected with METH in two environments that differed in ambient temperature (22–23 °C vs. 29 °C) while brain temperature and locomotion were recorded. The brains were taken at different times after METH administration, when NAc temperature peaked or reached clearly pathological values (>41.5 °C). The state of BBB permeability and edema were then determined by albumin immunoreactivity and the measuring of brain water content. Albumin is a relatively large plasma protein (molecular weight 59 kDa, molecular diameter 70 A) that is normally confined to the luminal side of endothelial cells and is not present in the brain under normal conditions. Thus, the appearance of albumin immunoreactivity in brain cells or neuropil indicates a breakdown of the BBB. Finally, brain slices were examined using light and electron microscopy to determine the extent and specifics of cellular abnormalities.

As shown in Fig. 2, METH induced a robust increase in albumin immunoreactivity, suggesting BBB leakage. Compared to saline-treated controls, albumin immunoreactivity increased strongly in both METH groups and the changes were significantly larger when the drug was administered at 29° than 23 °C (C). While rats in each condition showed significant increases in NAc temperature, the increases were significantly stronger when METH was administered at 29 °C (A). Both METH groups also showed significant increases in brain water content and this increase was significantly larger in rats that received METH at 29 °C (B). As shown in Fig. 3A, albumin immunoreactivity was strongly dependent on brain temperature, with virtually no positive cells appearing at low basal temperatures and a progressive increase occurring at high temperatures. Tissue water content was directly related to brain temperature (B), suggesting a tight relationship between brain hyperthermia and edema.

Fig. 4 shows leakage of serum albumin in the cerebral cortex in rats that were administered METH at 23 °C (B) and 29° (C) ambient temperatures. In contrast to the control (A), in METH-treated rats albumin-positive cells were found within the neuropil (arrows) largely in neurons. In rats exposed

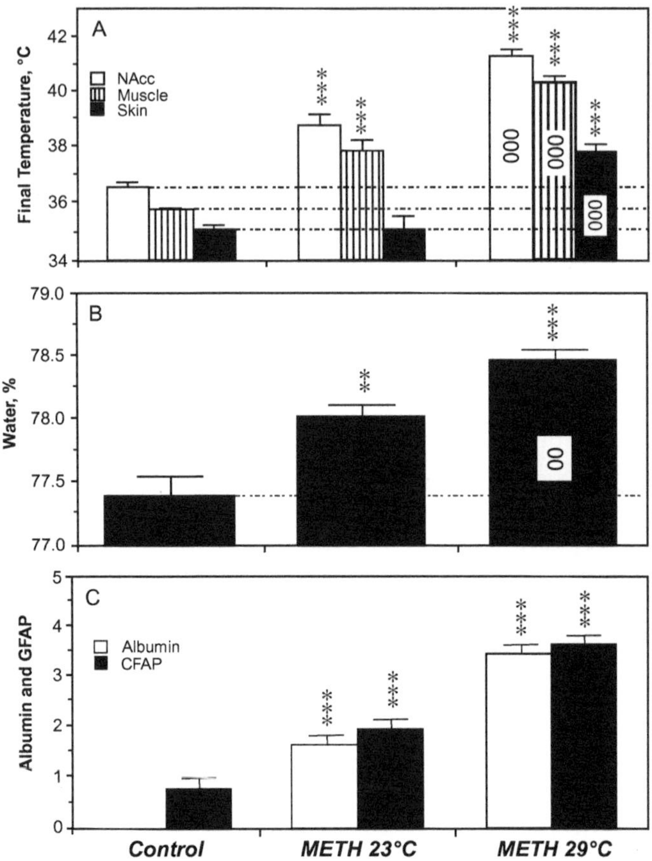

Fig. 2 Mean (±SEM) values of temperature (A), brain tissue water content (B) and immunoreactivity for albumin and GFAP (C) in rats exposed to METH in standard (23 °C) and warm (29 °C) environmental temperatures. Asterisks show values significantly different from the control (intact rats that received saline injections), and circles indicate significant differences between 23 and 29 °C. *Original data were published in Kiyatkin, E.A., Brown, P.L., Sharma, H.S. (2007). Brain edema and breakdown of blood-brain barrier during methamphetamine intoxication: Critical role of brain temperature.* The European Journal of Neuroscience 26, *1242–1253.*

to METH at 29 °C, the intensity of staining was stronger than that in the METH-23 °C group, with the frequent appearance of distorted neurons with perineuronal edema.

It appears that METH-induced leakage of the BBB and accumulation of brain tissue water are phenomena occurring in the brain as a whole. While METH increased albumin immunoreactivity and water tissue content in

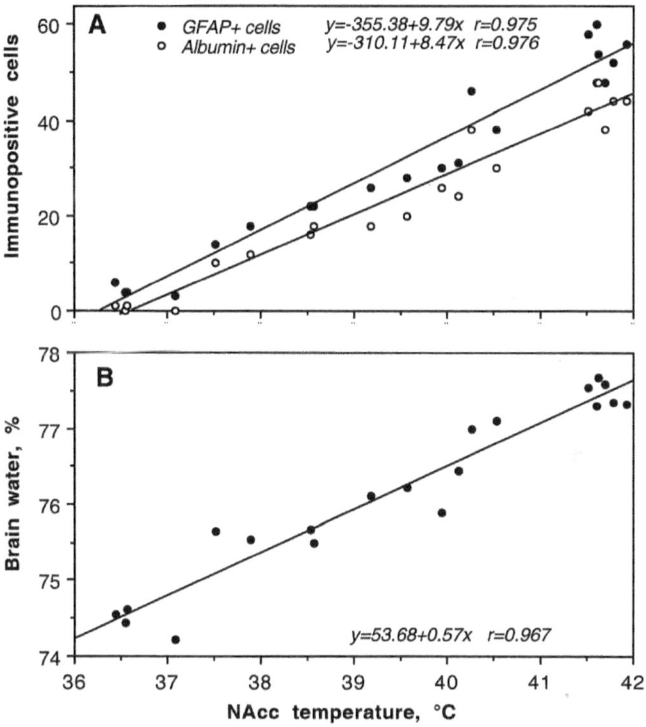

Fig. 3 Correlative relationships between METH-induced changes in brain temperature and (A) albumin and GFAP immunoreactivity and (B) brain tissue water content. The changes in all three neural parameters were tightly correlated with brain temperature (r = coefficient of correlation). *Original data were published in Kiyatkin, E.A., Brown, P.L., Sharma, H.S. (2007). Brain edema and breakdown of blood-brain barrier during methamphetamine intoxication: Critical role of brain temperature.* The European Journal of Neuroscience 26, *1242–1253.*

each of tested structures and the effect was significantly stronger when METH was administered at 29 °C, the extent of changes differed among individual structures (Fig. 5).

Acute METH exposure also induced structural abnormalities of brain cells (Fig. 6). The number of abnormal cells assessed in the cortex tightly correlated with brain and muscle temperatures (A), albumin leakage and glial activation (B), and tissue water accumulation (C). Morphological cell abnormalities were virtually absent in the saline-treated control, significantly increased in the METH-23 °C group, and were further increased in the METH-29 °C group. Correlations between all of these parameters were linear within the entire range of their fluctuations (Fig. 6).

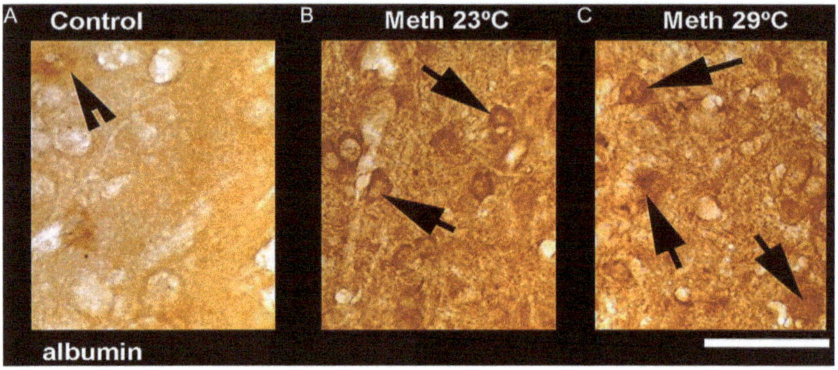

Fig. 4 Leakage of serum albumin in the cerebral cortex in rats exposed to METH at 23 °C (B) and 29 °C (C) as compared to the saline-treated control (A). In METH-treated rats, albumin positive cells are seen within the neuropil (arrows) largely in the neurons (b and c). Distorted neurons with perineuronal edema were more frequent in rats that received METH at 29 °C compared to 23 °C. *Data were modified after Sharma, H.S. and Kiyatkin, E.A. (2009). Rapid morphological brain abnormalities during acute methamphetamine intoxication in the rat: An experimental study using light and electron microscopy. Journal of Chemical Neuroanatomy 37, 18–32.*

Fig. 7 shows several examples of ultrastructural changes in the cortex (a, c, and e) and thalamus (b, d, and f) in rats exposed to METH at 23 °C (c and d) and 29 °C (e and f) as compared to saline-treated control (a, b). Degeneration of the neuronal nucleus and eccentric nucleolus (arrow) is clearly seen in METH treated rats; the magnitude and intensity of cellular and membrane degeneration is higher at 29 °C compared to 23 °C (a, c, and e). In the thalamus, myelin vesiculation (d and f, arrows), expansion of the neuropil (*) and membrane disruption were quite common in the METH-treated group. The intensity of these changes were aggravated much more by METH treatment at 29 °C (f) compared to the identical treatment at 23 °C (d).

Therefore, these data suggest that acute METH intoxication results in robust breakdown of the BBB, development of edema, and structural abnormalities in brain cells. These effects are enhanced when the drug is administered in a moderately warm environment, and they tightly correlate with drug-induced brain temperature elevations. Finally, the neurotoxic effects of METH depend not only on direct drug action on brain cells and drug-induced metabolic activation, but they are also strongly modulated by the environmental conditions associated with the drug administration.

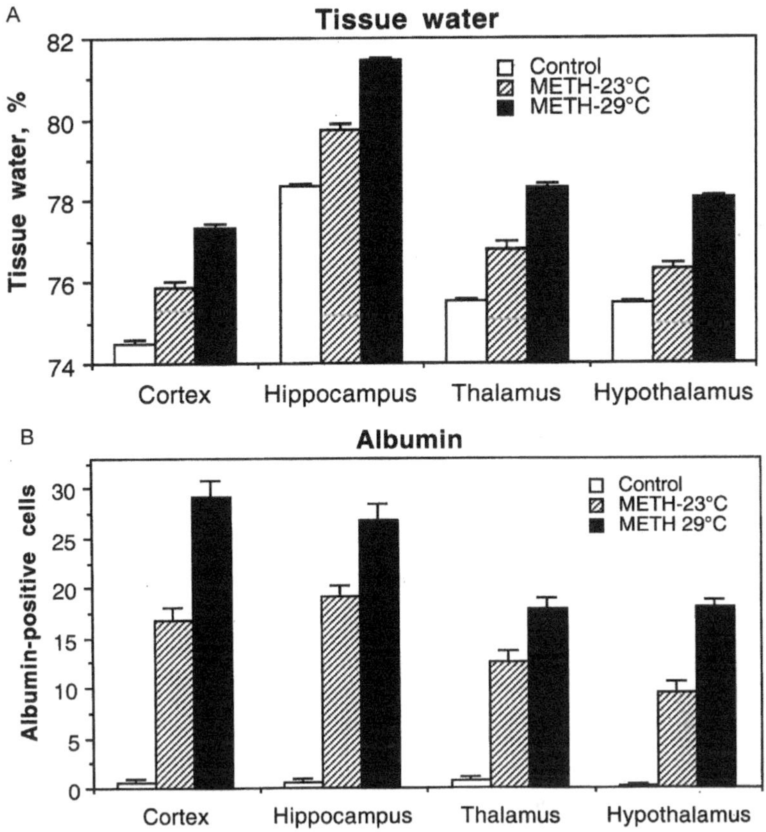

Fig. 5 Changes in brain tissue water (A) and albumin immunoreactivity (B) in several brain structures induced by METH administered to freely-moving rats at 23 °C and 29 °C ambient temperatures.

4. Conclusions and functional implications

In this work, we discussed the possible role of BBB leakage and subsequent changes in water homeostasis in acute METH toxicity and established the links between these changes and brain hyperthermia, possibly the most dangerous effect of this drug. While usually underappreciated, brain temperature is an important physiological parameter that reflects metabolic neural activity and affects all types of neural function. While the dose of an injected or ingested drug is usually viewed as the most important parameter for inducing pathological hyperthermia and overdose lethality,

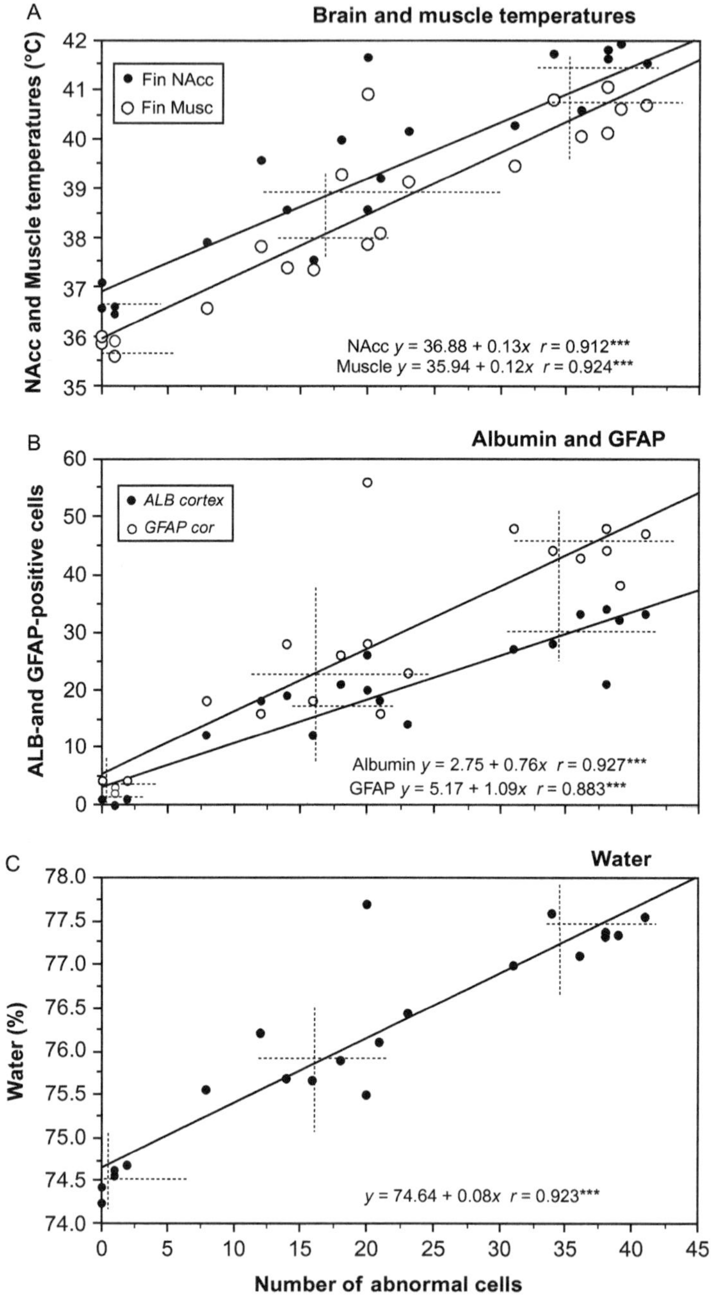

Fig. 6 Correlative relationships between METH-induced structural abnormalities of brain cells (x-axis) and: (A) brain and muscle temperatures; (B) albumin and GFAP immunoreactivity; and (C) water content in brain tissue. The changes in all neural parameters tightly correlated with brain temperature (r = coefficient of correlation). *Original data were published in Kiyatkin, E.A., Brown, P.L., Sharma, H.S. (2007). Brain edema and breakdown of blood-brain barrier during methamphetamine intoxication: Critical role of brain temperature.* The European Journal of Neuroscience 26, *1242–1253.*

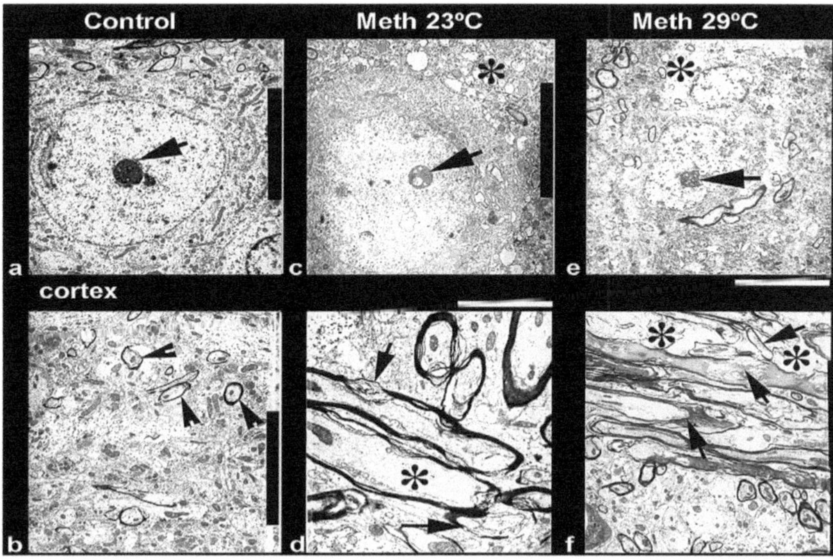

Fig. 7 Ultrastructural changes in the cortex (a, c, and e) and thalamus (b, d, and f) after METH treatment at 23 °C (c and d) and 29 °C (e and f) as compared to the control (a and b). Degeneration of the neuronal nucleus and eccentric nucleolus (arrow) is clearly seen in METH-treated rats, and the magnitude and intensity of cellular and membrane degeneration was higher at 29 °C as compared to 23 °C (a, c, and e). In the thalamus, myelin vesiculation (d and f, arrows), expansion of neuropil (*) and membrane disruption were quite common in the METH-treated group. The intensity of these changes was more aggravated by METH treatment at 29 °C (f) as compared to identical treatment at 23 °C (d). *Data are modified from Sharma, H.S. and Kiyatkin, E.A. (2009). Rapid morphological brain abnormalities during acute methamphetamine intoxication in the rat: An experimental study using light and electron microscopy. Journal of Chemical Neuroanatomy 37, 18–32.*

this review demonstrates that the temperature effects of drugs such as METH are strongly enhanced when they are used during physiological activation and in moderately warm environments. Social interaction, a procedure used in rats to model human psychophysiological activation, moderately enhanced intra-brain heat production and diminished heat dissipation due to peripheral vasoconstriction. The combination of these effects with the similar but more sustained effects induced by psychomotor stimulants results in a strong potentiation of drug-induced hyperthermic responses. A similar but stronger potentiation occurred when METH was administered in a warm environment where proper heat dissipation was important to maintain brain and body temperature homeostasis. Therefore, a certain drug used under adverse environmental conditions could induce unexpectedly strong effects and even death at doses that are usually viewed as "safe." We also show that METH intoxication results in strong leakage of the BBB and accumulation of brain tissue water indicating vasogenic edema. These changes could be responsible for multiple abnormalities of different types of brain cells. All of the parameters indicating structural abnormalities of brain cells were strongly increased in rats that received METH at warm temperature vs. normal ambient temperature. A strong correlation was also found between changes in these parameters in individual rats of both groups. While robust temperature increases and the possible action of potentially toxic products of brain metabolism could definitely contribute to acute toxicity of METH, the strength of drug-induced changes in water content suggests that this change could be the leading factor determining development of coma and lethality. While rats differ from humans in their ability to regulate brain temperature and any animal experiment cannot fully mimic human conditions, we believe that our findings could add something new in terms of understanding the pathophysiological mechanisms underlying acute METH toxicity and lethality during METH overdose.

Acknowledgments

The study was supported by the Intramural Research Program of the NIH, NIDA.

References

Abrams, R., & Hammel, H. T. (1964). Hypothalamic temperature in unanesthetized albino rats during feeding and sleeping. *The American Journal of Physiology*, *206*, 641–646.

Alberts, D. S., & Sonsalla, P. K. (1995). Methamphetamine-induced hyperthermia and dopaminergic neurotoxicity in mice: Pharmacological profile of protective and nonprotective agents. *The Journal of Pharmacology and Experimental Therapeutics*, *275*, 1104–1114.

Baker, M. A., Frye, F. M., & Millet, V. E. (1973). Origin of temperature changes evoked in the brain by sensory stimulation. *Experimental Neurology*, *38*, 502–519.

Banks, M. L., Sprague, J. E., Kisor, D. F., Czoty, P. W., Nichols, D. E., & Nader, M. A. (2007). Ambient temperature effects on 3,4-methylenedioxymethamphetamine-induced theremodysregulation and pharmacokinetics in male monkeys. *Drug Metabolism and Disposition*, *35*, 1840–1845.

Brown, P. L., Wise, R. A., & Kiyatkin, E. A. (2003). Brain hyperthermia is induced by methamphetamine and exacerbated by social interaction. *The Journal of Neuroscience*, *23*, 3924–3929.

Cadet, J. L., Krasnova, I. N., Jayanthi, S., & Lyles, J. (2007). Neurotoxicity of substituted amphetamines: Molecular and cellular mechanisms. *Neurotoxicity Research*, *11*, 183–202.

Cervos-Navarro, J., Sharma, H. S., Westman, J., & Bongcum-Rudloff, E. (1998). Glial cell reactions in the central nervous system following heat stress. *Progress in Brain Research*, *115*, 241–274.

Chen, Y. Z., Xu, R. X., Huang, Q. J., Xu, Z. J., Jiang, X. D., & Cai, Y. O. (2003). Effect of hyperthermia on tight junctions between endothelial cells of the blood-brain barrier model in vitro. *Di Yi Jun Da Xue Xue Bao*, *23*, 21–24.

Davis, W. M., Hatoum, H. T., & Walters, I. W. (1987). Toxicity of MDA (2.4-methylenedioxyamphetamine) considered for relevance to hazards of MDMA (Ecstasy) abuse. *Alcohol and Drug Research*, 7, 123–134.
De Vito, M. J., & Wagner, G. C. (1989). Methamphetamine-induced neuronal damage: A possible role for free radicals. *Neuropharmacology, 28*, 1145–1150.
Delgado, J. M. R., & Hanai, T. (1966). Intracerebral temperatures in free-moving cats. *The American Journal of Physiology, 211*, 755–769.
Esposito, P., Cheorghe, D., Kendere, K., Pang, X., Connoly, R., Jaconson, S., et al. (2001). Acute stress increases permeability of the blood-brain barrier through activation of brain must cells. *Brain Research, 888*, 117–127.
Freedman, R. R., Johnson, C. E., & Tancer, M. E. (2005). Thermoregulatory effects of 3,4-methylenedioxymethamphetamine (MDMA) in humans. *Psychophamacology, 183*, 248–256.
Fuller, C. A., & Baker, M. A. (1983). Selective regulation of brain and body temperatures in the squirrel monkey. *The American Journal of Physiology, 245*, R293–R297.
Gordon, C. J. (1990). Thermal physiology of the laboratory rat. *Physiology & Behavior, 47*, 963–991.
Gordon, C. J. (2007). Thermophysiological responses to hyperthermic drugs: Extrapolating from rodents to human. *Progress in Brain Research, 162*, 63–79.
Hayward, J. N., & Baker, M. A. (1968). Role of cerebral blood flow in the regulation of brain temperature in the monkey. *The American Journal of Physiology, 215*, 389–403.
Iwagami, Y. (1996). Changes in the ultrastructure of human cell related to certain biological responses under hyperthermic culture conditions. *Human Cell, 9*, 353–366.
Kalant, H. (2001). The pharmacology and toxicology of "ecstasy" (MDMA) and related drugs. *Canadian Medical Association Journal, 165*, 917–928.
Kiyatkin, E. A. (2018). Brain temperature: From physiology and pharmacology to neuropathology. *Handbook of Clinical Neurology, 157*, 483–504.
Kiyatkin, E. A., & Brown, P. L. (2005). Brain and body temperature homeostasis during sodium pentobarbital anesthesia with and without body warming in rats. *Physiology & Behavior, 84*, 563–570.
Kiyatkin, E. A., Brown, P. L., & Sharma, H. S. (2007). Brain edema and breakdown of blood-brain barrier during methamphetamine intoxication: Critical role of brain temperature. *The European Journal of Neuroscience, 26*, 1242–1253.
Kiyatkin, E. A., Brown, P. L., & Wise, R. A. (2002). 2002. Brain temperature fluctuation: A reflection of functional neural activation. *The European Journal of Neuroscience, 16*, 164–168.
Kiyatkin, E. A., & Mitchum, R. (2003). Fluctuations in brain temperatures during sexual behavior in male rats: An approach for evaluating neural activity underlying motivated behavior. *Neuroscience, 119*, 1169–1183.
Kovalson, V. M. (1972). Brain temperature variations during natural sleep and arousal in white rats. *Physiology & Behavior, 10*, 667–670.
Lepock, J. R. (2003). Cellular effects of hyperthermia: Relevance to the minimum dose for thermal damage. *International Journal of Hyperthermia, 19*, 252–266.
Lynch, J., & House, M. A. (1992). Cardiovascular effects of methamphetamine. *The Journal of Cardiovascular Nursing, 6*, 12–18.
McElligott, J. C., & Melzack, R. (1967). Localized thermal changes evoked in the brain by visual and auditory stimulation. *Experimental Neurology, 17*, 293–312.
Miller, D. B., & O'Callaghan, J. P. (2003). Elevated environmental temperature and methamphetamine neurotoxicity. *Environmental Research, 92*, 48–53.
Nybo, L., Secher, M. H., & Nielson, B. (2002). Inadequate heat release from the human brain during prolonged exercise with hyperthermia. *The Journal of Physiology, 545*, 697–704.

Oifa, A. I., & Kleshchnov, V. N. (1985). Ultrastructural analysis of the phenomenon of acute neuronal swelling. *Zhurnal nevropatologii i psikhiatrii imeni S.S. Korsakova, 85*, 1016–1020.

Ovadia, H., Abramsky, O., Feldman, S., & Weidenfeld, J. (2001). Evaluation of the effects of stress on the blood-brain barrier: Critical role of the brain perfusion time. *Brain Research, 905*, 21–25.

Prakash, M. D., Tanalakis, K., Antonipillal, J., Stojanovska, L., Nurdall, K., & Apostolopoulos, V. (2017). Methamphetamine: Effects on the brain, gut, and immune system. *Pharmacological Research, 120*, 60–67.

Rapoport, S. I. (1976). *Blood-brain barrier in physiology and medicine*. New York: Raven Press.

Ricaurte, G. A., Schuster, C. R., & Seiden, L. S. (1980). Long-term effects of repeated methamphetamine administration on dopamine and serotonin neurons in the rat brain: A regional study. *Brain Research, 193*, 153–163.

Riddle, E. L., Fleckenstein, A. E., & Hanson, G. R. (2006). Mechanisms of methamphetamine-induced dopaminergic neurotoxicity. *The AAPS Journal, 8*(2), Article 48. http://www.aapsj.org.

Romanovsky, A. A., Ivanov, A. I., & Shimansky, Y. P. (2002). Ambient temperature for experiments in rats: A new method for determining the zone of thermal neutrality. *Journal of Applied Physiology, 92*, 2667–2679.

Rowell, L. B. (1983). Cardiovascular aspects of human thermoregulation. *Circulation Research, 52*, 367–379.

Rusyniak, D. E., & Sprague, J. E. (2005). Toxin-induced hyperthermic syndromes. *The Medical Clinics of North America, 89*, 1277–1296.

Sandoval, V., Hanson, G. R., & Fleckenstein, A. E. (2000). Methamphetamine decreases mouse striatal dopamine transport activity: Roles of hyperthermia and dopamine. *European Journal of Pharmacology, 409*, 265–271.

Schmidt-Nielsen, K. (1997). Animal physiology. In *Adaptation and environment*. (5th ed.). Cambridge: Cambridge University Press.

Seiden, L. S., & Sabol, K. E. (1996). Methamphetamine and methylenedioxymethamphetamine neurotoxicity: Possible mechanisms of cell destruction. *NIDA Research Monograph, 163*, 251–276.

Serota, H. M., & Gerard, R. M. (1938). Localized thermal changes in cat's brain. *Journal of Neurophysiology, 1*, 115–124.

Sharma, H. S., & Ali, S. F. (2006). Alterations in blood-brain barrier function by morphine and amphetamine. *Annals of the New York Academy of Sciences, 1074*, 198–224.

Sharma, H. S., & Dey, P. K. (1986). Influence of long-term immobilization stress on regional blood-brain permeability, cerebral blood flow and 5-HT levels in conscious normotensive young rats. *Journal of the Neurological Sciences, 72*, 61–76.

Sharma, H. S., & Hoopes, P. J. (2003). Hyperthermia-induced pathophysiology of the central nervous system. *International Journal of Hyperthermia, 19*, 325–354.

Sharma, H. S., & Kiyatkin, E. A. (2009). Rapid morphological brain abnormalities during acute methamphetamine intoxication in the rat: An experimental study using light and electron microscopy. *Journal of Chemical Neuroanatomy, 37*, 18–32.

Sharma, H. S., Zimmer, C., Westman, J., & Cervos-Navarro, J. (1992). Acute systemic heat stress increases glial fibrillary acidic protein immunoreactivity in brain. An experimental study in the conscious normotensive young rats. *Neuroscience, 48*, 889–901.

Siesjo, B. (1978). *Brain energy metabolism*. New York: Wiley.

Smirnov, M. S., & Kiyatkin, E. A. (2008). Fluctuations in central and peripheral temperatures associated with feeding behavior in rats. *The American Journal of Physiology, 295*, R1414–R1424.

Stephans, S. E., & Yamamoto, B. K. (1994). Methamphetamine-induced neurotoxicity: Role for glutamate and dopamine influx. *Synapse, 17*, 203–209.

Taffe, M. A., Lay, C. C., Von Huben, S. N., Davis, S. A., Crean, R. D., & Katner, S. N. (2006). Hyperthermia induced by 3,4-methylenedioxymethamphetamine in unrestrained rhesus monkeys. *Drug and Alcohol Dependence, 20*, 276–281.

Von Huben, R. D., Lay, C. C., Crean, R. D., Davis, S. A., Katner, S. N., & Taffe, M. A. (2007). Impact of ambient temperature on hyperthermia induced by (+/−)3,4-methylenediomethamphetamine in rhesus macaques. *Neuropsychopharmacology, 32*, 673–681.

Watson, P., Shirreffs, S. M., & Maughan, R. J. (2005). Blood-brain barrier integrity may be threatened by exercise in a warm environment. *The American Journal of Physiology, 288*, R1689–R1694.

Woolverton, W. L., Ricaurte, G. A., Forno, L., & Seiden, L. S. (1989). Long term effects of chronic methamphetamine administration in rhesus monkeys. *Brain Research, 486*, 73–78.

Yamamoto, H. (1963). The central effects of xylopinine in mice. *Japanese journal of pharmacology, 13*, 230–239.

Yamamoto, B. K., & Zhu, W. (1998). The effect of methamphetamine on the production of free radicals and oxidative stress. *The Journal of Pharmacology and Experimental Therapeutics, 287*, 107–114.

Zlokovic, B. V. (2008). The blood-brain barrier in health and chronic neurodegenerative disorders. *Neuron, 57*, 178–201, 2008.

Further reading

O'Callaghan, J. P. (1993). Quantitative features of reactive gliolis following toxicant-induced damage of the CNS. *Annals of the New York Academy of Sciences, 679*, 195–210.

Sharma, H. S. (2006). Hyperthermia-induced brain edema: Current status and future perspectives. *The Indian Journal of Medical Research, 123*, 629–652.

CHAPTER FIVE

Therapeutic potential of cannabinoids as neuroprotective agents for damaged cells conducing to movement disorders

Mario Antonazzo[a,b], María Botta[c], Harkaitz Bengoetxea[d], José Ángel Ruiz-Ortega[a,b,c], Teresa Morera-Herreras[a,b,*]

[a]Department of Pharmacology, Faculty of Medicine and Nursing, University of the Basque Country (UPV/EHU), Leioa, Spain
[b]Neurodegenerative Diseases Group, BioCruces Bizkaia Health Research Institute, Barakaldo, Bizkaia, Spain
[c]Department of Pharmacology, Faculty of Pharmacy, University of the Basque Country (UPV/EHU), Vitoria-Gasteiz, Spain
[d]Department of Neurosciences, Faculty of Medicine and Nursing, University of the Basque Country (UPV/EHU), Leioa, Spain
*Corresponding author: e-mail address: teresa.morera@ehu.eus

Contents

Abstract

The basal ganglia (BG), an organized network of nuclei that integrates cortical information, play a crucial role in controlling motor function. In fact, movement disorders such as Parkinson's disease (PD) and Huntington's disease (HD) are caused by the degeneration of specific structures within the BG. There is substantial evidence supporting the idea that cannabinoids may constitute novel promising compounds for the treatment of movement disorders as neuroprotective and anti-inflammatory agents. This potential therapeutic role of cannabinoids is based, among other qualities, on their capacity to reduce oxidative injury and excitotoxicity, control calcium influx and limit the toxicity of reactive microglia. The mechanisms involved in these effects are related to CB1 and CB2 receptor activation, although some of the effects are CB receptor independent.

International Review of Neurobiology, Volume 146
ISSN 0074-7742
https://doi.org/10.1016/bs.irn.2019.06.012

Thus, taking into account the aforementioned properties, compounds that act on the endocannabinoid system could be useful as a basis for developing disease-modifying therapies for PD and HD.

Abbreviations

2-AG	2-arachidonoyl glycerol
3-NP	3-nitropropionic acid
6-OHDA	6-hydroxydopamine
Δ^9-THC	Δ^9-tetrahydrocannabinol
ACEA	arachidonyl-2′-chloroethylamide
AEA	arachidonoyl ethanolamide
BG	basal ganglia
CAG	cytosine-adenine-guanine
CBD	cannabidiol
CNS	central nervous system
COX-2	cyclooxygenase 2
DAGLα	diacylglycerol lipase α
ECS	endocannabinoid system
FAAH	fatty acid amide hydrolase
GABA	gamma aminobutyric acid
HD	Huntington's disease
HTT	Huntingtin protein
IL	interleukin
iNOS	inducible nitric oxide synthase
LPS	lipopolysaccharide
MAGL	monoacylglycerol lipase
MPTP	1-methyl-4-phenyl-1,2,3,6-tetrahydropyridine
MSN	medium spiny neuron
NAPE-PLD	N-acyl-phosphatidylethanolamine-hydrolysing phospholipase D
NO	nitric oxide
PPAR	peroxisome proliferator-activated receptor
PD	Parkinson's disease
ROS	reactive oxygen species
SNpc	*substantia nigra pars compacta*
SNpr	*substantia nigra pars reticulata*
TNF-α	tumor necrosis factor-α
TRPV1	transient receptor potential vanilloid type 1
VTA	ventral tegmental area

1. The endocannabinoid signaling system and its distribution in the basal ganglia

Preparations of the cannabis plant have been used as medicinal drugs by various cultures and civilizations for thousands of years. The identification of more than 60 compounds known as cannabinoids that are present in the plant *Cannabis sativa* led to the investigation and discovery of the endocannabinoid system (ECS). The ECS comprises lipid-signaling endogenous ligands named endocannabinoids (eCBs), biosynthesizing and degrading enzymes, and their associated receptors and is broadly distributed in the central and peripheral nervous systems (Castillo, Younts, Chávez, & Hashimotodani, 2012).

The best-characterized eCBs in the central nervous system (CNS) are 2-arachidonoyl glycerol (2-AG) and arachidonoyl ethanolamide (AEA, also known as anandamide) (Devane et al., 1992; Sugiura, Kishimoto, Oka, & Gokoh, 2006). The main degradative enzyme for 2-AG in the brain, monoacylglycerol lipase (MAGL), is located presynaptically (Gulyas et al., 2004) and is the crucial element for the control of 2-AG signaling (Hashimotodani, Ohno-Shosaku, & Kano, 2007). In contrast, diacylglycerol lipase α (DAGLα) is localized in postsynaptic neurons (Katona & Freund, 2008), and it is required for Ca^{2+}-dependent 2-AG production and eCB-mediated synaptic plasticity (Tanimura et al., 2010). Postsynaptic depolarization and intracellular Ca^{2+} influx also promote AEA production. One of the main enzymes that synthesizes AEA in the brain is N-acyl-phosphatidylethanolamine-hydrolysing phospholipase D (NAPE-PLD), which can be expressed both post- and presynaptically (Cristino et al., 2008), whereas fatty acid amide hydrolase (FAAH) is mainly implicated in AEA catabolism (Varvel, Cravatt, Engram, & Lichtman, 2006).

The effects of eCBs are primarily mediated by their interaction with two well-characterized $G_{i/0}$ protein-coupled cannabinoid receptor subtypes, CB1 and CB2 (Castillo et al., 2012). Activation of these receptor subtypes inhibits adenylyl cyclase and some voltage-dependent calcium channels and activates mitogen-activated protein kinases (MAPK) and inwardly rectifying potassium channels. In the brain, the concentration of 2-AG is higher than that of AEA, and 2-AG acts as a full agonist, whereas AEA acts as a partial agonist at cannabinoid receptors, suggesting that cannabinoid regulation in the brain is elicited mainly by 2-AG (Sugiura et al., 2006). In addition, eCBs can also interact with other receptors such as the transient receptor potential vanilloid type 1 (TRPV1) cation channel, the GTP-binding protein-coupled receptor GPR55, the abnormal-CBD receptor and the peroxisome proliferator-activated receptor (PPAR).

CB1 and CB2 receptors are widely expressed throughout the body, and countless brain functions, including anxiety, mood, cognition, motor control, feeding behavior and pain, as well as pathophysiological processes such as Parkinson's disease (PD) and dyskinesias, are regulated at least in part through these receptors. The CB1 receptor resides in axon terminal endings, but substantial functional evidence confirms that it is also expressed on somata (Fitzgerald, Shobin, & Pickel, 2012). This receptor is highly expressed in the CNS, including the cortex, cerebellum, hippocampus and especially the basal ganglia (BG). The BG, an organized network of subcortical nuclei that connects the cortex and the thalamus, display the highest levels of both mRNA expression and CB1 receptor (Romero et al., 2002). The striatum (Calabresi, Picconi, Tozzi, Ghiglieri, & Di Filippo, 2014), as well as other regions such as the *globus pallidus*, the *entopeduncular nucleus* and the *substantia nigra pars reticulata* (SNpr), which receive striatal efferent outputs, have high levels of CB1 receptor binding sites (Muñoz-Arenas et al., 2015; Sierra et al., 2015). CB1 receptors are presynaptically located in striatonigral (direct striatal efferent pathway) and striatopallidal (indirect striatal efferent pathway) projection neurons (Kofalvi, 2005). These neurons use gamma-aminobutyric acid (GABA) as a neurotransmitter. Other markers that are co-expressed with GABA in these pathways include glutamic acid decarboxylase, prodynorphin, substance P, preproenkephalin, and D1 and D2 dopaminergic receptors (Hohmann & Herkenham, 2000). In contrast, intrinsic striatal neurons, which contain acetylcholine or somatostatin, do not express CB1 receptors (Hohmann & Herkenham, 2000). Moreover, CB1 receptors have been observed in glutamatergic corticostriatal afferents (Gerdeman & Lovinger, 2001; Kofalvi, 2005) and in subthalamopallidal and/or subthalamonigral projections (Dasilva, Grieve, Cudeiro, & Rivadulla, 2014; Mailleux & Vanderhaeghen, 1992). Apart from neuronal populations, CB1 receptors are also expressed on glial cells, but their density is lower than that observed on neurons (Bilkei-Gorzo et al., 2018).

On the other hand, CB2 receptors are more predominant in peripheral tissues and cells of the immune system (B lymphocytes, natural killer cells, monocytes, neutrophils, T8 lymphocytes, and T4 lymphocytes) (Munro, Thomas, & Abu-Shaar, 1993; Navarrete, García-Gutiérrez, Aracil-Fernández, Lanciego, & Manzanares, 2018). CB2 receptors are also expressed in the CNS, especially in the dorsal vagal motor nucleus, the nucleus ambiguus and the spinal trigeminal nucleus (Núñez et al., 2008; Van Sickle et al., 2005). In addition, although early attempts to identify CB2 receptors in the CNS under basal conditions failed, this paradigm has changed in the last decade. For the first time, Van Sickle et al. (2005) revealed that CB2 receptors are expressed in neurons of the brainstem of mice, rats and ferrets under normal conditions. Following this pioneering work, subsequent studies have characterized the expression of CB2 receptors in several brain regions, including the cortex, BG (striatum, SNpr), amygdala, hippocampus and ventral tegmental area (VTA) (Gong et al., 2006; Onaivi et al., 2006; Zhang et al., 2014). In these brain regions, CB2 receptors were detected in microglia (Cabral, Raborn, Griffin, Dennis, & Marciano-Cabral, 2008) and in neurons (Zhang et al., 2014). Thus, the presence of CB2 receptors has been described in cortical, hippocampal, pallidal and mesencephalic neurons (Lanciego et al., 2011) as well as in the amygdala and striatum (Brusco, Tagliaferro, Saez, & Onaivi, 2008; Onaivi et al., 2006). Because other studies did not localize these receptors in healthy brains, the localization of CB2 receptors in the CNS remains controversial. The disagreement among authors on this point may be due in part to the lack of specificity of the antibodies used. To circumvent this methodological limitation, new genetic strategies based on mouse lines that express enhanced green fluorescent protein under the control of the CB2 promoter have been developed (López et al., 2018).

Various pharmacological, neurochemical and electrophysiological studies indicate that eCBs function largely through retrograde signaling in which postsynaptic activity induces eCB production and release with backward transmission across the synapse to depress presynaptic neurotransmitter release (Castillo et al., 2012). In the BG, eCBs tend to decrease GABAergic and glutamatergic transmission by activating CB1 receptors (Benarroch, 2007). Moreover, eCBs interfere with the dopaminergic system (reviewed in Morera-Herreras, Miguelez, Aristieta, Ruiz-Ortega, & Ugedo, 2012), affecting dopamine release and modulating the firing activity of dopaminergic neurons (Solinas, Justinova, Goldberg, & Tanda, 2006). Although nigrostriatal dopaminergic neurons seem not to express CB1 receptors (Piomelli, 2003), the activation or the blockade of the ECS directly affects the functionality of these cells (Fernández-Ruiz, Hernández, & Ramos, 2010). These actions are mediated by the CB1 receptor, which is expressed in neuronal subpopulations that are located in the vicinity and are connected to dopaminergic neurons such as GABAergic, glutamatergic, and opioidergic neurons (van der Stelt & Di Marzo, 2003). In addition, the stimulation of dopamine receptors increases the levels of striatal AEA, which may serve as an inhibitory feedback signal countering dopamine-induced motor activity (Ferrer, Asbrock, Kathuria, Piomelli, & Giuffrida, 2003). However, additional mechanisms could explain the modulation of dopaminergic transmission by eCBs, since some eCBs interact with other receptors such as TRPV1, which is expressed on nigrostriatal terminals and on tyrosine hydroxylase-positive neurons in the *substantia nigra pars compacta* (SNpc) (Mezey et al., 2000; Starowicz, Nigam, & Di Marzo, 2007).

2. Mechanisms underlying neuroprotection by cannabinoids

The ECS is targeted by compounds that have specific biological and therapeutic effects; these effects include, among others, limitation of neuronal damage by reduction of excitotoxicity, neuroinflammation, oxidative stress and calcium influx. Accordingly, cannabinoids are emerging as potential neuroprotective compounds in progressive neurodegenerative pathologies (Fernández-Ruiz, Romero, & Ramos, 2015). In fact, pharmacological manipulation of the ECS has revealed neuroprotective effects in some experimental models of neurodegenerative diseases, suggesting that this therapeutic strategy could be useful in modifying the progression of such pathologies (Fernández-Ruiz et al., 2010).

The molecular mechanisms underlying the neuroprotective action of cannabinoids are diverse (summarized in the Fig. 1). On one hand, such neuroprotection is related to a direct action on the ECS by activation of CB1 and CB2 (Ramirez, 2005), TRPV1 (Baek et al., 2018) or GPR55 receptors (Kallendrusch et al., 2013). On the other hand, neuroprotection by cannabinoids could also be mediated by activating targets outside the ECS, such as nuclear receptors of the PPAR family (Fidaleo, Fanelli, Ceru, & Moreno, 2014), transcription factors (Fernández-Ruiz et al., 2013), serotonin 1A receptors (Pazos et al., 2013) and the adenosine signaling pathway (Castillo, Tolón, Fernández-Ruiz, Romero, & Martinez-Orgado, 2010). The broad spectrum of activity of cannabinoid compounds may be important in its use as a neuroprotective agent since multi-targeted designed drugs have been shown to be very promising tools in preclinical studies testing neuroprotective and disease-modifying agents (Geldenhuys & Van der Schyf, 2013).

2.1 Neuroprotection directly related to CB receptors

Substantial evidence supports the idea that excitotoxicity plays a key role in neuronal damage in neurodegenerative diseases. In this sense, CB1 receptors can provide a neuroprotective effect by normalizing glutamate homeostasis. CB1 receptors are located at neuronal glutamatergic terminals, and their activation by cannabinoid agonists reduces glutamate release, thereby limiting glutamatergic excitotoxicity (Mechoulam, Panikashvili, & Shohami, 2002; van der Stelt et al., 2002). In parallel with their antiglutamatergic effect, cannabinoids, by reducing the activation of NMDA receptors and the subsequent opening of voltage-sensitive calcium channels, can diminish calcium influx, thereby avoiding the activation of destructive pathways linked to high intracellular levels of this ion (Demuth & Molleman, 2006; Fowler et al., 2003).

Furthermore, the activation of CB1 receptors provides neuroprotection by inducing an enhancement of the blood supply after neuronal damage, an effect that could be of special interest in ischemic episodes. The mechanism implicated in this effect is related to a decrease in the generation of endothelium-derived mediators (endothelin-1 or nitric oxide (NO)) that induce vasoconstriction of injured tissues after ischemia (Chen et al., 2000).

In addition to their aforementioned beneficial effects, cannabinoids are also able to regulate the immunological processes that are clearly involved in neuronal damage, including local inflammatory events related to the activation of glial cells. In fact, many studies suggest that glial activation, along with neuroinflammation (microglial activation, astrogliosis or infiltration of immune cells), contributes to the initiation or progression of neurodegenerative processes. The activation of CB1 receptors leads to modulation of the levels of neurotrophins and anti-inflammatory mediators. Thus, cannabinoid ligands acting on CB1 receptors can create favorable conditions for the regeneration of damaged neurons by decreasing tumor necrosis factor-α (TNF-α) and interleukin-12 (IL-12) levels and increasing IL-10 levels (Smith, Terminelli, & Denhardt, 2000). Together with CB1 receptors, CB2 receptors are also implicated in these neuroprotective effects, since their activation inhibits the production of chemokines by astrocytes (Sheng, Hu, Ni, Rock, & Peterson, 2009). Moreover, the activation of CB2 receptors reduces the proliferation and migration of microglial cells at lesion sites (Walter et al., 2003) and reduces the production of TNF-α by inhibiting the activation of NF-κB (Oh et al., 2010), a transcription factor that is important in regulating pro-inflammatory responses.

Apart from the classical CB receptors, recent studies suggest that TRPV1 receptors may be a promising neuroprotective target. Capsaicin is able to protect neurons against neurotoxicity by specific activation of astrocytic TRPV1 receptors, which leads to reduced microglial activation and reduces consequent oxidative stress (Baek et al., 2018). In contrast, other authors have shown that overactivation of TRPV1 leads to neurotoxicity by increasing intracellular calcium levels or disrupting mitochondrial function (Kim et al., 2005). Therefore, further investigations are necessary to clarify the role of TRPV1 in neurodegeneration/neuroprotection.

Although its function in the brain is still poorly understood, the GPR55 receptor represents a putative receptor for various cannabinoids and has been proposed as a novel neuroprotective target. Thus, GPR55 receptor activation by lysophosphatidylinositol exerts a neuroprotective effect in excitotoxically lesioned rat organotypic cultures. However, the mechanisms behind GPR55 activation in microglia have yet to be unraveled (Kallendrusch et al., 2013).

2.2 Neuroprotection not directly related to CB receptors

Due to their antioxidant properties, cannabinoids can provide neuroprotection through mechanisms that are not directly related to their action at classical CB receptors. Phytocannabinoids such as Δ^9-tetrahydrocannabinol (Δ^9-THC), cannabinol and cannabidiol (CBD), as well as synthetic cannabinoids such as nabilone and HU-211, may protect against brain damage by restoring the equilibrium between oxidative and antioxidant mechanisms that is altered during neurodegenerative processes. In fact, during degeneration, a deficiency in endogenous antioxidant mechanisms and high levels of reactive oxygen species (ROS) have been described (Marsicano, Moosmann, Hermann, Lutz, & Behl, 2002). This neuroprotective potential of cannabinoids is related to their ability to reduce excessive production of these ROS by acting as scavengers and antagonizing lipid peroxidation (Chen & Buck, 2000; Hampson, Grimaldi, Axelrod, & Wink, 1998).

Further studies attempting to elucidate the exact mechanism underlying these antioxidant effects have suggested that, in addition to their particular chemical structure containing phenolic groups, the mechanism of action of cannabinoids may also be related to the activation of PPAR receptors. In fact, CBD exerts both antioxidant and anti-inflammatory effects through the activation of these receptors (Iuvone, Esposito, De Filippis, Scuderi, & Steardo, 2009; O'Sullivan, 2016). In particular, CBD inhibits the synthesis and release of a number of pro-inflammatory molecules, including cytokines, NO, and GFAP protein, from activated glial cells (Esposito et al., 2007) through a PPARγ receptor-mediated mechanism (Scuderi, Steardo, & Esposito, 2014). This effect is associated with the inhibition of p38 MAP kinase and the regulation of NF-κB, which controls the expression of genes that encode pro-inflammatory enzymes

(Esposito et al., 2006). Moreover, CBD regulates the microglial cell migration that plays a crucial role in the propagation of the neuroinflammatory response, preventing the recruitment of microglial cells to lesion sites (Walter et al., 2003). Nevertheless, it is known that CBD can enhance endogenous AEA signaling by inhibiting its intracellular degradation by FAAH or by decreasing eCB transport (Bisogno et al., 2001; Leweke et al., 2012). Consequently, CBD could provide neuroprotection by increasing eCB effects that are mediated by CB1 and CB2 receptor activation.

In addition, HU-211 is also able to act on postsynaptic NMDA glutamatergic receptors; it may decrease the intracellular concentration of calcium by closing voltage-dependent calcium channels, thereby also promoting a neuroprotective effect that is not mediated by CB receptor activation (Hansen et al., 2002; Nadler, Mechoulam, & Sokolovsky, 1993).

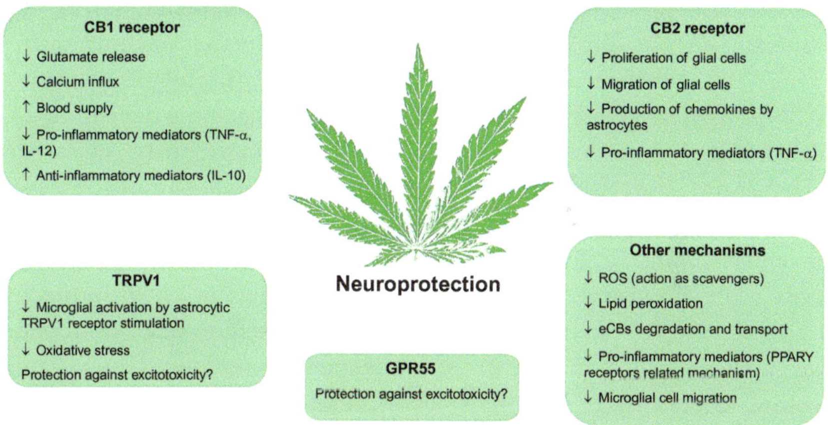

Fig. 1 Summary of the different mechanisms proposed for the neuroprotection exerted by cannabinoids. Abbreviations: eCBs, endocannabinoids; IL, interleukin; PPAR, peroxisome proliferator-activated receptor; ROS, Reactive oxygen species; TNF, tumor necrosis factor; TRPV1, transient receptor potential vanilloid type 1.

3. Cannabinoids as neuroprotective agents in neurodegenerative disorders

Due to the aforementioned pharmacological properties of the cannabinoids, compounds that target the ECS could be useful as neuroprotective tools in neurodegenerative disorders such as PD or Huntington's disease (HD) in which specific structures within the BG degenerate. Current pharmacological treatments for both pathologies focus on restoring normal motor functionality and ameliorating symptoms but are not able to arrest or slow down the progression of neurodegeneration. Therefore, the search for disease-modifying therapies is of primary interest in the treatment of PD and HD.

3.1 Neuroprotective effect of cannabinoids in Parkinson's disease

PD is the second most common neurodegenerative disorder after Alzheimer's disease. It is an age-related pathology that is present in 1–2% of the population over 60 years of age (Massano & Bhatia, 2012; Nussbaum & Ellis, 2003). Although the most notorious symptoms of PD include motor alterations such as bradykinesia, resting tremors, rigidity and postural instability, non-motor alterations (e.g., dementia and depression) are also present (reviewed in Poewe & Mahlknecht, 2010). The two main physiopathological hallmarks of PD are the accumulation of α-synuclein aggregates and the degeneration of dopaminergic neurons in the SNpc (reviewed in Thomas, 2009), leading to dopamine depletion. This will affect the functionality of the BG, key nuclei in voluntary movement and other processes, as they are strongly modulated by dopamine (reviewed in Gerfen & Surmeier, 2011). In recent decades, there has been an increase in our understanding of the pathological pathways that underlie PD development. This knowledge has suggested several targets for neuroprotective intervention in PD such as neuroinflammation, excitotoxicity, mitochondrial function and oxidative stress, among others (reviewed in Schapira, Olanow, Greenamyre, & Bezard, 2014). The ECS has been suggested as an appropriate target system for increasing neuroprotection because it is able to modulate some of the pathological features observed in PD patients' brains and in animal models (reviewed in Fernández-Ruiz et al., 2011) (summarized in Table 1).

Neuroinflammation occurs mainly through glial activation, which is thought to be a key contributor to dopaminergic degeneration (Kaur, Gill, Bansal, & Deshmukh, 2017; Lull & Block, 2010). Glial activation occurs both in patients and in animal models of PD and is due to the subsequent increase in the levels of various pro-inflammatory agents such as IL-1β, TNF-α, inducible nitric oxide synthase (iNOS) and cyclooxygenase 2 (COX-2) (Ouchi et al., 2005; Ramsey & Tansey, 2014; Teismann & Schulz, 2004).

Table 1 Neuroprotective effects of the endocannabinoid system modulation in Parkinson's disease.
Parkinson's disease

1. Reduction of neuroinflammation	
CB1 activation	↓ Microglial activation ↓ Production of pro-inflammatory agents (IL-1β, TNF-α, iNOs, COX-2)
CB1 blockade	↑ Activated astrocytes in SNpc
CB2 activation	↓ Glial cell number ↓ Production of pro-inflammatory agents (IL-1β, TNF-α, iNOs, NO)
MAGL inhibition	↑ Glial cell number ↑ Levels of GDNF ↑ Production of anti-inflammatory mediators (TGF-β)
2. Reduction of excitotoxicity	
CB1 activation	↓ Excitotoxicity by inhibiting glutamate release
FAAH inhibition	Prevention against excitotoxicity
3. Modulation of mitochondrial function	
PPAR-γ activation	↑ Mitochondrial biogenesis
CB1 activation	↑ Mitochondrial biogenesis ↑ Mitochondrial function ↑ Number of mitochondria
4. Reduction of oxidative stress	
CB1 activation	↓ Expresión of microglial NADPH oxidase in an MPTP animal model ↓ ROS ↓ Lipid peroxidation
CB2 activation	↓ Myeloperoxidase-positive astrocytes ↑ Antioxidant enzyme activity ↑ Antioxidant agents
PPAR-γ activation	↓ ROS
FAAH inhibition	↓ ROS ↓ Lipid peroxidation ↓ Protein carbonylation

Abbreviations: COX-2, cyclooxygenase 2; FAAH, fatty acid amide hydrolase; GDNF, Glial cell-derived neurotrophic factor; IL, interleukin; iNOS, inducible nitric oxide synthase; MAGL, monoacylglycerol lipase; MPTP, 1-methyl-4-phenyl-1,2,3,6-tetrahydropyridine; NO, nitric oxide; PPAR, peroxisome proliferator-activated receptor; ROS, reactive oxygen species; SNpc, substantia nigra pars compacta; TGF, Transforming growth factor; TNF, tumor necrosis factor

Several reports show the promising ability of cannabinoids to reduce neuroinflammation in PD models. *In vivo* CB1 activation diminished dopaminergic neuronal damage caused by the neurotoxin 1-methyl-1,2,3, 6-tetrahydropyridine (MPTP), decreased microglial activation, and reduced IL-1β, TNF-α and iNOS levels (Chung et al., 2011). However, other reports showed that these protective effects of cannabinoids against MPTP toxicity were independent of CB1 receptor activation (Price et al., 2009). Arachidonyl-2′-chloroethylamide (ACEA), a CB1-selective agonist, prevented neuronal death in an *in vitro* lipopolysaccharide (LPS)-induced inflammation model (Vrechi, Crunfli, Costa, & Torrão, 2018). In another report, rimonabant, a CB1 antagonist, increased the survival of SNpc dopaminergic neurons after striatal injection of 6-hydroxydopamine (6-OHDA). It was shown to have no effect on microglial activation but to increase the number of activated astrocytes in the SNpc (Cerri et al., 2014). The CB2 receptor agonist JWH-133 prevented MPTP-induced neurodegeneration and reduced the production of IL-1β, TNF-α and iNOS by activated microglia (Chung et al., 2016). Other publications also provide evidence for a role of the CB2 receptor in preventing neurodegeneration, reducing levels of inflammatory factors and decreasing glial cell numbers in animal models of PD (Javed, Azimullah, Haque, & Ojha, 2016; Ojha, Javed, Azimullah, & Haque, 2016) and in decreasing inflammatory markers in an LPS-induced inflammation model *in vitro* (Malek, Popiolek-Barczyk, Mika, Przewlocka, & Starowicz, 2015). In addition to receptors, eCB metabolic enzymes may also have potential as therapeutic targets. The MAGL inhibitor JZL184 partially prevented MPTP-induced neuronal damage *in vivo*; it also increased glial cell numbers and levels of GDNF and TGF-β, an anti-inflammatory cytokine (Fernández-Suárez et al., 2014). Moreover, cannabinoids had no neuroprotective effect in *in vitro* PD models when glia were not present, suggesting that their modulatory effect on glial activity is the main mechanism responsible for the neuroprotection offered by these compounds (Chung et al., 2011; Lastres-Becker, Molina-Holgado, Ramos, Mechoulam, & Fernández-Ruiz, 2005).

Neuroinflammation can contribute to glutamate release and thereby promote glutamate-related processes. Moreover, aggregated α-synuclein present in the brains of PD patients is known to increase glutamate-induced synaptic currents. Thus, excitotoxicity occurs due to the overstimulation of glutamatergic receptors. Furthermore, activation of glutamate receptors located on microglia causes the release of pro-inflammatory cytokines, feeding neuroinflammation, glutamate-related excitotoxicity, and subsequent neuronal loss. Although not the reason for the initial neuronal decay, this process contributes to dopaminergic degeneration in PD (reviewed in Ambrosi, Cerri, & Blandini, 2014).

As mentioned above, cannabinoids can regulate glutamate release, thus providing a way to reduce excitotoxicity (Domenici, 2006; Takahashi & Castillo, 2006). Δ^9-THC and AEA were shown to reduce the excitotoxicity induced by ouabain *in vivo* through activation of the CB1 receptor. In the same experimental model, modulation of TPRV1 activity also reduced excitotoxicity (van der Stelt et al., 2001; Veldhuis et al., 2003). Moreover, WIN 55,212-2 and AEA, CB1/CB2 receptor agonists, prevented neuronal death in *in vitro* models of excitotoxicity (Rangel-López et al., 2015; Serpa, Pinto, Bernardino, & Cascalheira, 2015). Other reports have suggested that BDNF might be a key to CB1-mediated protection against excitotoxicity (Blázquez et al., 2015; Khaspekov et al., 2004). Regulation of eCB metabolism might also be useful in fighting excitotoxicity in PD, considering that FAAH inhibition was shown to prevent excitotoxicity in various excitotoxic models both *in vivo* and *in vitro* (Aguilera-Portillo et al., 2019; Karanian et al., 2007; Naidoo et al., 2011).

Cannabinoids can also modulate mitochondrial function and, therefore, oxidative stress through a variety of mechanisms; thus, they represent potential therapeutic tools for reducing such processes in PD. PD patients display mitochondrial dysfunction that is reflected in reduced activity of complex I of the mitochondrial respiratory electron transport chain. Moreover, some complex I inhibitors have been used to recreate PD in animal models, and some mutations linked to PD are involved in mitochondrial function. This mitochondrial dysfunction is linked to ROS production, which causes further mitochondrial and cellular damage leading to apoptosis and thus contributes to neuronal degeneration in PD (reviewed in Schapira, 2008).

Some cannabinoids may have positive effects on mitochondrial biogenesis and function. For instance, Δ^9-THC is able to restore the levels of proteins involved in mitochondrial biogenesis and protect cells from MPTP toxicity *in vitro* in a PPAR-γ-dependent manner (Carroll, Zeissler, Hanemann, & Zajicek, 2012; Zeissler et al., 2016). In *in vivo* and *in vitro* models of cerebral ischemia, activation of the CB1 receptor was shown to increase the number of mitochondria as well as to enhance the expression of markers of mitochondrial biogenesis and function (Bai et al., 2017; Cai et al., 2017; Ma et al., 2015). Furthermore, CBD reversed the dopaminergic impairment caused by 6-OHDA injection in an animal model (Lastres-Becker et al., 2005) and had positive effects on mitochondrial function and oxidative stress in *in vitro* models of various pathological states (Liu et al., 2014; Sun, Hu, Wu, & Zhang, 2017; Yang et al., 2014). One clinical trial investigated the possible neuroprotective effects of CBD in PD patients and found no differences in the neuroprotection parameters that were assessed (Chagas et al., 2014). The limitations of this study were its small sample size and, possibly, the short treatment duration used to evaluate changes in PD progression.

Finally, oxidative stress induced *in vivo* in a rotenone PD model was reduced by administration of *Cannabis sativa* extracts, although no protection against dopaminergic degeneration was observed (Abdel-Salam et al., 2015). Both CB1 and CB2 receptors seem to be involved in the antioxidant activity of cannabinoids. WIN 55,212-2 offered protection against neurodegeneration, reduced lipid peroxidation and protein carbonylation *in vivo* after MPTP treatment (Escamilla-Ramírez et al., 2017), and reduced ROS and lipid peroxidation *in vitro* after quinolinic acid insult (Rangel-López et al., 2015). Activation of the CB1 receptor resulted in reduced expression of microglial NADPH oxidase and decreased DNA and protein oxidative damage in an MPTP animal model of PD (Chung et al., 2011). CB2 receptor activation reduced the number of myeloperoxidase-positive astrocytes and increased antioxidant enzyme activity levels and the levels of antioxidant molecules in animal models of PD (Chung et al., 2016; Javed et al., 2016). Δ^9-THC also prevented dopaminergic lesions after 6-OHDA injection *in vivo* (Lastres-Becker et al., 2005) and reduced ROS production in an *in vitro* model of PD through PPARγ activation (Carroll et al., 2012; Zeissler et al., 2016). URB597, a FAAH inhibitor, also exerted neuroprotective effects against neuronal death induced by 6-OHDA and MPTP *in vivo* by reducing lipid peroxidation, protein carbonylation and ROS (Escamilla-Ramírez et al., 2017; Maya-López, Ruiz-Contreras, et al., 2017; Pelição et al., 2016). Moreover, some cannabinoids have intrinsic structural antioxidant properties that also contribute to their effectiveness as antioxidant agents (Marsicano et al., 2002).

3.2 Neuroprotective effect of cannabinoids in Huntington's disease

HD is a rare autosomal-dominant progressive neurodegenerative disorder caused by a mutation in the IT15 gene (*HTT*) that encodes the huntingtin protein (HTT). The mutation consists of an extension of the trinucleotide cytosine-adenine-guanine (CAG) repeat (>39 CAG repeats) that yields an abnormally long polyglutamine sequence in the N-terminus of HTT (pathogenic HTT) (MacDonald et al., 1993). The neuropathological consequence of this mutation is the specific degeneration of striatal GABAergic medium spiny neurons (MSNs) (Reiner et al., 1988) and, to a lesser extent, of glutamatergic pyramidal neurons in the cerebral cortex (Hedreen, Peyser, Folstein, & Ross, 1991; Heinsen et al., 1994). The neurodegeneration leads to motor abnormalities, the early stages of which are characterized by chronic movements and cognitive deterioration; the advanced stages are associated with parkinsonian-like symptoms such as bradykinesia and rigidity (Berardelli et al., 1999). Although the molecular mechanisms underlying neuronal degeneration and disease onset in HD have not been elucidated in detail, some authors have demonstrated that mitochondrial dysfunction, energy depletion, excitotoxicity, oxidative damage, protein aggregation and neuroinflammation play important roles in the neuronal loss that occurs in HD (for review, see Borrell-Pagès, Zala, Humbert, & Saudou, 2006).

There is substantial evidence supporting the idea that the ECS in the BG becomes hypofunctional in HD. In fact, progressive loss of CB1 receptors in murine animal models of HD has been reported (Bisogno, Martire, Petrosino, Popoli, & Di Marzo, 2008; Dowie et al., 2009; Glass, Van Dellen, Blakemore, Hannan, & Faull, 2004; McCaw et al., 2004). These observations are consistent with *post mortem* autoradiographic studies of brain tissue sections from HD patients that show a notable decrease in CB1 receptor levels in MSNs of the caudate and putamen at early stages of the disease (Denovan-Wright & Robertson, 2000; Glass, Dragunow, & Faull, 2000; Van Laere et al., 2010). This could very likely be due to inhibition of CB1 transcription mediated by the mutant HTT (Laprairie et al., 2014). Interestingly, a massive and progressive loss of CB1 receptors occurs prior to the neurodegeneration of MSNs and the onset of motor symptoms (Dowie et al., 2009; McCaw et al., 2004). These data suggest that activation of CB1 receptors may modify the progression of HD.

The hypothesis that CB1 receptors may modify the progression of HD has been intensively investigated, mainly in preclinical experimental models of the disease. In these models, most of the neuroprotective actions observed were mediated via CB1 cannabinoid receptors. Because these receptors are located both presynaptically and postsynaptically at neuronal glutamatergic terminals, their activation by cannabinoid agonists would be expected to lead to decreased glutamate release and thus to limit excitotoxicity (Sagredo et al., 2007). In fact, administration of the CB1 receptor agonist WIN 55,212-2 to an intrastriatal quinolinic acid-injected rat model of HD preserved striatal neurons by decreasing glutamate-mediated

Table 2 Neuroprotective effects of the endocannabinoid system modulation in Huntington's disease
Huntington's disease

1. Reduction of excitotoxicity	
CB1 activation	↓Glutamate-mediated excitotoxicity
2. Attenuation of glial toxicity	
CB2 activation	↓Reactive microglial cells ↓Production of pro-inflammatory cytokines (TNF-α) ↓ROS ↓NO ↑Production of neurotrophins and anti-inflammatory mediators (IL-10, IL-1 antagonist)
3. Antioxidant/antiinflammatory properties	
Chemical structure	↓ROS (action as scavengers)
PPAR-γ activation	Interference with the NF-κB signaling pathway Induction of antioxidant enzymes

Abbreviations: IL, interleukin; NF-κB, nuclear factor-κB; NO, nitric oxide; PPAR, peroxisome proliferator-activated receptor; ROS, reactive oxygen species; TNF, tumor necrosis factor

excitotoxicity (Table 2). This effect was blocked by the CB1 receptor antagonist AM251, demonstrating that it was CB1-dependent (Pintor et al., 2006). Similarly, administration of Δ^9-THC or WIN 55,212-2 had a neuroprotective effect against striatal neurotoxicity in rats lesioned with the mitochondrial toxin 3-nitropropionic acid (3-NP) (Lastres-Becker et al., 2004; Maya-López, Colín-González, et al., 2017). Using the same HD model in mice, Díaz-Alonso et al. (2016) found that a novel cannabigerol derivative, VCE-003.2, also prevented loss of MSNs. Additionally, in the R6/2 mutant mouse model, activation of CB1 receptors by the administration of Δ^9-THC also protected MSNs from death (Blázquez et al., 2011). In contrast, in double-mutant animals (CB1 receptor-deficient R6/2 mutant mice), striatal degeneration was exacerbated due to increased excitotoxic damage and decreased levels of brain-derived neurotrophic factors (Blázquez et al., 2011; Chiarlone et al., 2014; Mievis, Blum, & Ledent, 2011). Rescue of CB1 gene expression by striatal injection of a recombinant adeno-associated viral vector encoding the CB1 receptor normalized the HD pathology characteristic of this mouse model (Blázquez et al., 2015; Naydenov et al., 2014).

Overall, the foregoing evidence indicates that CB1 receptor stimulation could be a promising neuroprotective tool for delaying HD progression. Although substantial loss of CB1 receptor-containing striatal neurons occurs in advanced stages of the disease (Fernández-Ruiz et al., 2015), it was recently demonstrated that the neuroprotective effect of such stimulation is predominantly due to the specific activation of CB1 receptors located on corticostriatal terminals, which are preserved during the progression of the disease (Chiarlone et al., 2014).

Cannabinoid compounds have also been shown to exert neuroprotective effects in HD by selectively activating CB2 receptors (Table 2). This cannabinoid receptor subtype is poorly abundant in the striatum in conditions of health, but its overexpression has been described both in HD patients and in animal models of the disease (Fernández-Ruiz, Pazos, García-Arencibia, Sagredo, & Ramos, 2008; Fernández-Ruiz et al., 2007; Palazuelos et al., 2009). Upregulation of CB2 receptors occurs in glial cells, including astrocytes, and is especially marked in striatal reactive microglia (Fernández-Ruiz et al., 2010, 2007). CB2 receptors may control the influence of reactive microglia on neurons by reducing the release of pro-inflammatory cytokines, ROS or NO and, moreover, by promoting the production of neurotrophins and anti-inflammatory mediators (IL-10 or IL-1 receptor antagonists) (reviewed in Navarro et al., 2016). In this way, the compensatory overexpression of CB2 receptors exerts a beneficial effect in HD, protecting striatal neurons from neuroinflammatory injury. In fact, genetic ablation of the CB2 receptor in R6/2 mice exacerbates microglial activation, aggravates motor abnormalities and accelerates the onset of the disease (Bouchard et al., 2012; Palazuelos et al., 2009). Consistent with these findings, selective pharmacological stimulation of CB2 receptors has a neuroprotective effect on striatal neurons in various animal models of HD such as quinolinic acid-lesioned mice and malonate-lesioned rats (Palazuelos et al., 2009; Sagredo et al., 2009).

Finally, cannabinoids such as Δ^9-THC and CBD have been shown to exert neuroprotective action in the 3-NP lesioned rat model of HD due to antioxidant mechanisms that are independent of CB1/CB2 receptor activation (Sagredo et al., 2007, 2011) (Table 2). These effects seem to be related to a scavenging action on ROS that is probably due to the phenolic structures of these phytocannabinoids. In addition, recent evidence demonstrates that cannabinoids also produce beneficial effects by activating PPAR-γ in transgenic R6/2 and 3-NP-lesioned mice (reviewed in Fernández-Ruiz, Gómez-Ruiz, García, Hernández, & Ramos, 2017).

Despite the vast amount of preclinical evidence indicating that cannabinoids may be a promising neuroprotective agent in HD, the clinical studies conducted to date have focused on evaluating the potential role of cannabinoids in the alleviation of specific symptoms without determining their effect on disease progression. In a double-blind, randomized, crossover, placebo-controlled pilot clinical trial, administration of Sativex® to 25 HD patients for 12 weeks failed to improve symptomatology, produce molecular changes in biomarkers or slow disease progression (López-Sendón Moreno et al., 2016). However, in a more recent study, seven HD patients showed improved dystonia and motor scores after cannabinoid drug administration (Sativex®, nabilone or dronabinol) (Saft et al., 2018). In these studies, cannabinoids were safe and well-tolerated and were associated with a low incidence of adverse effects. Additional clinical trials are warranted to assess the potential value of cannabinoid-based therapies in halting or slowing the progression of neurodegeneration in HD. Nevertheless, the authors suggest that in future studies, higher doses and/or longer treatment periods will be needed.

4. Concluding remarks

The solid preclinical evidence that has been presented to date demonstrates the neuroprotective properties of cannabinoid compounds and shows that these compounds represent promising drugs for delaying the progression of neurodegenerative diseases associated with cytotoxicity-induced neuronal and glial damage. The neuroprotective "multi-target" approach offered by cannabinoids must now be evaluated and validated in clinical trials. Such trials could offer a framework for the development of novel disease-modifying therapies in the treatment of PD or HD.

References

Abdel-Salam, O. M. E., Youness, E. R., Khadrawy, Y. A., Mohammed, N. A., Abdel-Rahman, R. F., Omara, E. A., et al. (2015). The effect of cannabis on oxidative stress and neurodegeneration induced by intrastriatal rotenone injection in rats. *Comparative Clinical Pathology*, *24*(2), 359–378. https://doi.org/10.1007/s00580-014-1907-9.

Aguilera-Portillo, G., Rangel-López, E., Villeda-Hernández, J., Chavarría, A., Castellanos, P., Elmazoglu, Z., et al. (2019). The pharmacological inhibition of fatty acid amide hydrolase prevents excitotoxic damage in the rat striatum: Possible involvement of CB1 receptors regulation. *Molecular Neurobiology*, *56*, 844–856. https://doi.org/10.1007/s12035-018-1129-2.

Ambrosi, G., Cerri, S., & Blandini, F. (2014). A further update on the role of excitotoxicity in the pathogenesis of Parkinson's disease. *Journal of Neural Transmission*, *121*(8), 849–859. https://doi.org/10.1007/s00702-013-1149-z.

Baek, J. Y., Jeong, J. Y., Kim, K. I., Won, S.-Y., Chung, Y. C., Nam, et al. (2018). Inhibition of microglia-derived oxidative stress by ciliary neurotrophic factor protects dopamine neurons in vivo from MPP+ neurotoxicity. *International Journal of Molecular Sciences*, *19*(11), 3543. https://doi.org/10.3390/ijms19113543.

Bai, F., Guo, F., Jiang, T., Wei, H., Zhou, H., Yin, H., et al. (2017). Arachidonyl-2-Chloroethylamide alleviates cerebral ischemia injury through glycogen synthase kinase-3β-mediated mitochondrial biogenesis and functional improvement. *Molecular Neurobiology*, *54*(2), 1240–1253. https://doi.org/10.1007/s12035-016-9731-7.

Benarroch, E. (2007). Endocannabinoids in basal ganglia circuits: Implications for Parkinson disease. *Neurology*, *69*(3), 306–309. https://doi.org/10.1212/01.wnl.0000267407.79757.75.

Berardelli, A., Noth, J., Thompson, P. D., Bollen, E. L. E. M., Currà, A., Deuschl, G., et al. (1999). Pathophysiology of chorea and bradykinesia in Huntington's disease. *Movement Disorders*, *14*(3), 398–403. https://doi.org/10.3989/cyv.2007.v46.i1.260.

Bilkei-Gorzo, A., Albayram, O., Ativie, F., Chasan, S., Zimmer, T., Bach, K., et al. (2018). Cannabinoid 1 receptor signaling on GABAergic neurons influences astrocytes in the ageing brain. *PLoS One*, *13*(8), e0202566. https://doi.org/10.1371/journal.pone.0202566.

Bisogno, T., Hanuš, L., De Petrocellis, L., Tchilibon, S., Ponde, D. E., Brandi, I., et al. (2001). Molecular targets for cannabidiol and its synthetic analogues: Effect on vanilloid VR1 receptors and on the cellular uptake and enzymatic hydrolysis of anandamide. *British Journal of Pharmacology*, *134*(4), 845–852. https://doi.org/10.1038/sj.bjp.0704327.

Bisogno, T., Martire, A., Petrosino, S., Popoli, P., & Di Marzo, V. (2008). Symptom-related changes of endocannabinoid and palmitoylethanolamide levels in brain areas of R6/2 mice, a transgenic model of Huntington's disease. *Neurochemistry International*, *52*(1), 307–313. https://doi.org/10.1016/j.neuint.2007.06.031.

Blázquez, C., Chiarlone, A., Bellocchio, L., Resel, E., Pruunsild, P., García-Rincón, D., et al. (2015). The CB1cannabinoid receptor signals striatal neuroprotection via a PI3K/Akt/mTORC1/BDNF pathway. *Cell Death and Differentiation*, *22*(10), 1618–1629. https://doi.org/10.1038/cdd.2015.11.

Blázquez, C., Chiarlone, A., Sagredo, O., Aguado, T., Pazos, M. R., Resel, E., et al. (2011). Loss of striatal type 1 cannabinoid receptors is a key pathogenic factor in Huntington's disease. *Brain*, *134*(1), 119–136. https://doi.org/10.1093/brain/awq278.

Borrell-Pagès, M., Zala, D., Humbert, S., & Saudou, F. (2006). Huntington's disease: From huntingtin function and dysfunction to therapeutic strategies. *Cellular and Molecular Life Sciences*, *63*(22), 2642–2660. https://doi.org/10.1007/s00018-006-6242-0.

Bouchard, J., Truong, J., Bouchard, K., Dunkelberger, D., Desrayaud, S., Moussaoui, S., et al. (2012). Cannabinoid receptor 2 signaling in peripheral immune cells modulates disease onset and severity in mouse models of Huntington's disease. *Journal of Neuroscience*, *32*(50), 18259–18268. https://doi.org/10.1523/JNEUROSCI.4008-12.2012.

Brusco, A., Tagliaferro, P., Saez, T., & Onaivi, E. S. (2008). Postsynaptic localization of CB2 cannabinoid receptors in the rat hippocampus. *Synapse*, *62*(12), 944–949. https://doi.org/10.1002/syn.20569.

Cabral, G. A., Raborn, E. S., Griffin, L., Dennis, J., & Marciano-Cabral, F. (2008). CB 2 receptors in the brain: Role in central immune function. *British Journal of Pharmacology*, *153*(2), 240–251. https://doi.org/10.1038/sj.bjp.0707584.

Cai, M., Yang, Q., Li, G., Sun, S., Chen, Y., Tian, L., et al. (2017). Activation of cannabinoid receptor 1 is involved in protection against mitochondrial dysfunction and cerebral ischaemic tolerance induced by isoflurane preconditioning. *British Journal of Anaesthesia*, *119*(6), 1213–1223. https://doi.org/10.1093/bja/aex267.

Calabresi, P., Picconi, B., Tozzi, A., Ghiglieri, V., & Di Filippo, M. (2014). Direct and indirect pathways of basal ganglia: A critical reappraisal. *Nature Neuroscience*, *17*(8), 1022–1030. https://doi.org/10.1038/nn.3743.

Carroll, C. B., Zeissler, M. L., Hanemann, C. O., & Zajicek, J. P. (2012). Δ9-tetrahydrocannabinol (Δ9-THC) exerts a direct neuroprotective effect in a human cell culture model of Parkinson's disease. *Neuropathology and Applied Neurobiology*, *38*(6), 535–547. https://doi.org/10.1111/j.1365-2990.2011.01248.x.

Castillo, A., Tolón, M. R., Fernández-Ruiz, J., Romero, J., & Martinez-Orgado, J. (2010). The neuroprotective effect of cannabidiol in an in vitro model of newborn hypoxic-ischemic brain damage in mice is mediated by CB2and adenosine receptors. *Neurobiology of Disease*, *37*(2), 434–440. https://doi.org/10.1016/j.nbd.2009.10.023.

Castillo, P. E., Younts, T. J., Chávez, A. E., & Hashimotodani, Y. (2012). Endocannabinoid signaling and synaptic function. *Neuron*, *76*(1), 70–81. https://doi.org/10.1016/j.neuron.2012.09.020.

Cerri, S., Levandis, G., Ambrosi, G., Montepeloso, E., Antoninetti, G. F., Franco, R., et al. (2014). Neuroprotective potential of adenosine A2Aand cannabinoid CB1receptor antagonists in an animal model of Parkinson disease. *Journal of Neuropathology and Experimental Neurology*, *73*(5), 414–424. https://doi.org/10.1097/NEN.0000000000000064.

Chagas, M. H. N., Zuardi, A. W., Tumas, V., Pena-Pereira, M. A., Sobreira, E. T., Bergamaschi, M. M., et al. (2014). Effects of cannabidiol in the treatment of patients with Parkinson's disease: An exploratory double-blind trial. *Journal of Psychopharmacology*, *28*(11), 1088–1092. https://doi.org/10.1177/0269881114550355.

Chen, Y., & Buck, J. (2000). Cannabinoids protect cells from oxidative cell death: A receptor-independent mechanism. *The Journal of Pharmacology and Experimental Therapeutics*, *293*(3), 807–812. https://doi.org/10.1016/0306-4573(95)80035-R.

Chen, Y., McCarron, R. M., Ohara, Y., Bembry, J., Azzam, N., Lenz, F. A., et al. (2000). Human brain capillary endothelium: 2-Arachidonoglycerol (endocannabinoid) interacts with endothelin-1. *Circulation Research*, *87*(4), 323–327. https://doi.org/10.1161/01.RES.87.4.323.

Chiarlone, A., Bellocchio, L., Blazquez, C., Resel, E., Soria-Gomez, E., Cannich, A., et al. (2014). A restricted population of CB1 cannabinoid receptors with neuroprotective activity. *Proceedings of the National Academy of Sciences of the United States of America*, *111*(22), 8257–8262. https://doi.org/10.1073/pnas.1400988111.

Chung, Y. C., Bok, E., Huh, S. H., Park, J.-Y., Yoon, S.-H., Kim, S. R., et al. (2011). Cannabinoid receptor type 1 protects nigrostriatal dopaminergic neurons against MPTP neurotoxicity by inhibiting microglial activation. *The Journal of Immunology*, *187*(12), 6508–6517. https://doi.org/10.4049/jimmunol.1102435.

Chung, Y. C., Shin, W. H., Baek, J. Y., Cho, E. J., Baik, H. H., Kim, S. R., et al. (2016). CB2 receptor activation prevents glial-derived neurotoxic mediator production, BBB leakage and peripheral immune cell infiltration and rescues dopamine neurons in the MPTP model of Parkinson's disease. *Experimental and Molecular Medicine*, *48*(1), e205. https://doi.org/10.1038/emm.2015.100.

Cristino, L., Starowicz, K., De Petrocellis, L., Morishita, J., Ueda, N., Guglielmotti, V., et al. (2008). Immunohistochemical localization of anabolic and catabolic enzymes for anandamide and other putative endovanilloids in the hippocampus and cerebellar cortex of the mouse brain. *Neuroscience*, *151*(4), 955–968. https://doi.org/10.1016/j.neuroscience.2007.11.047.

Dasilva, M., Grieve, K. L., Cudeiro, J., & Rivadulla, C. (2014). Anandamide activation of CB1 receptors increases spontaneous bursting and oscillatory activity in the thalamus. *Neuroscience*, *265*, 72–82. https://doi.org/10.1016/j.neuroscience.2014.01.049.

Demuth, D. G., & Molleman, A. (2006). Cannabinoid signalling. *Life Sciences*, *78*, 549–563. https://doi.org/10.1016/j.lfs.2005.05.055.

Denovan-Wright, E. M., & Robertson, H. A. (2000). Cannabinoid receptor messenger RNA levels decrease in a subset of neurons of the lateral striatum, cortex and hippocampus of transgenic Huntington's disease mice. *Neuroscience*, *98*(4), 705–713. https://doi.org/10.1016/S0306-4522(00)00157-3.

Devane, W. A., Hanuš, L., Breuer, A., Pertwee, R. G., Stevenson, L. A., Griffin, G., et al. (1992). Isolation and structure of a brain constituent that binds to the cannabinoid receptor. *Science, 258*(5090), 1946–1949. https://doi.org/10.1126/science.1470919.

Díaz-Alonso, J., Paraíso-Luna, J., Navarrete, C., Del Río, C., Cantarero, I., Palomares, B., et al. (2016). VCE-003.2, a novel cannabigerol derivative, enhances neuronal progenitor cell survival and alleviates symptomatology in murine models of Huntington's disease. *Scientific Reports, 6*(1), 29789. https://doi.org/10.1038/srep29789.

Domenici, M. R. (2006). Cannabinoid receptor type 1 located on presynaptic terminals of principal neurons in the forebrain controls glutamatergic synaptic transmission. *Journal of Neuroscience, 26*(21), 5794–5799. https://doi.org/10.1523/JNEUROSCI.0372-06.2006.

Dowie, M. J., Bradshaw, H. B., Howard, M. L., Nicholson, L. F. B., Faull, R. L. M., Hannan, A. J., et al. (2009). Altered CB1 receptor and endocannabinoid levels precede motor symptom onset in a transgenic mouse model of Huntington's disease. *Neuroscience, 163*(1), 456–465. https://doi.org/10.1016/j.neuroscience.2009.06.014.

Escamilla-Ramírez, A., García, E., Palencia-Hernández, G., Colín-González, A. L., Galván-Arzate, S., Túnez, I., et al. (2017). URB597 and the cannabinoid WIN55,212-2 reduce behavioral and neurochemical deficits induced by MPTP in mice: Possible role of redox modulation and NMDA receptors. *Neurotoxicity Research, 31*(4), 532–544. https://doi.org/10.1007/s12640-016-9698-1.

Esposito, G., De Filippis, D., Maiuri, M. C., De Stefano, D., Carnuccio, R., & Iuvone, T. (2006). Cannabidiol inhibits inducible nitric oxide synthase protein expression and nitric oxide production in β-amyloid stimulated PC12 neurons through p38 MAP kinase and NF-κB involvement. *Neuroscience Letters, 399*(1–2), 91–95. https://doi.org/10.1016/j.neulet.2006.01.047.

Esposito, G., Scuderi, C., Savani, C., Steardo, L., De Filippis, D., Cottone, P., et al. (2007). Cannabidiol in vivo blunts β-amyloid induced neuroinflammation by suppressing IL-1β and iNOS expression. *British Journal of Pharmacology, 151*(8), 1272–1279. https://doi.org/10.1038/sj.bjp.0707337.

Fernández-Ruiz, J., Gómez-Ruiz, M., García, C., Hernández, M., & Ramos, J. A. (2017). Modeling neurodegenerative disorders for developing cannabinoid-based neuroprotective therapies. In *Vol. 593. Methods in enzymology* (pp. 175–198). Academic Press. https://doi.org/10.1016/bs.mie.2017.06.021.

Fernández-Ruiz, J., Hernández, M., & Ramos, J. A. (2010). Cannabinoid-dopamine interaction in the pathophysiology and treatment of CNS disorders. *CNS Neuroscience & Therapeutics, 16*(3), e72–e91. https://doi.org/10.1111/j.1755-5949.2010.00144.x.

Fernández-Ruiz, J., Moreno-Martet, M., Rodríguez-Cueto, C., Palomo-Garo, C., Gomez-Cañas, M., Valdeolivas, S., et al. (2011). Prospects for cannabinoid therapies in basal ganglia disorders. *British Journal of Pharmacology, 163*(7), 1365–1378. https://doi.org/10.1111/j.1476-5381.2011.01365.x.

Fernández-Ruiz, J., Pazos, M. R., García-Arencibia, M., Sagredo, O., & Ramos, J. A. (2008). Role of CB2receptors in neuroprotective effects of cannabinoids. *Molecular and Cellular Endocrinology, 286*(1–2 Suppl. 1), S91–S96. https://doi.org/10.1016/j.mce.2008.01.001.

Fernández-Ruiz, J., Romero, J., & Ramos, J. A. (2015). Endocannabinoids and neurodegenerative disorders: Parkinson's disease, Huntington's chorea, Alzheimer's disease, and others. In R. G. Pertwee (Ed.), *Endocannabinoids* (pp. 233–259). Cham: Springer International Publishing. https://doi.org/10.1007/978-3-319-20825-1_8.

Fernández-Ruiz, J., Romero, J., Velasco, G., Tolón, R. M., Ramos, J. A., & Guzmán, M. (2007). Cannabinoid CB2receptor: A new target for controlling neural cell survival? *Trends in Pharmacological Sciences, 28*(1), 39–45. https://doi.org/10.1002/aqc.967.

Fernández-Ruiz, J., Sagredo, O., Pazos, M. R., García, C., Pertwee, R., Mechoulam, R., et al. (2013). Cannabidiol for neurodegenerative disorders: Important new clinical applications for this phytocannabinoid? *British Journal of Clinical Pharmacology, 75*(2), 323–333. https://doi.org/10.1111/j.1365-2125.2012.04341.x.

Fernández-Suárez, D., Celorrio, M., Riezu-Boj, J. I., Ugarte, A., Pacheco, R., González, H., et al. (2014). The monoacylglycerol lipase inhibitor JZL184 is neuroprotective and alters glial cell phenotype in the chronic MPTP mouse model. *Neurobiology of Aging, 35*(11), 2603–2616. https://doi.org/10.1016/j.neurobiolaging.2014.05.021.

Ferrer, B., Asbrock, N., Kathuria, S., Piomelli, D., & Giuffrida, A. (2003). Effects of levodopa on endocannabinoid levels in rat basal ganglia: Implications for the treatment of levodopa-induced dyskinesias. *European Journal of Neuroscience, 18*(6), 1607–1614. https://doi.org/10.1046/j.1460-9568.2003.02896.x.

Fidaleo, M., Fanelli, F., Ceru, M., & Moreno, S. (2014). Neuroprotective properties of peroxisome proliferator-activated receptor alpha (PPARα) and its lipid ligands. *Current Medicinal Chemistry, 21*(24), 2803–2821. https://doi.org/10.2174/0929867321666140303143455.

Fitzgerald, M. L., Shobin, E., & Pickel, V. M. (2012). Cannabinoid modulation of the dopaminergic circuitry: Implications for limbic and striatal output. *Progress in Neuro-Psychopharmacology and Biological Psychiatry, 38*(1), 21–29. https://doi.org/10.1016/j.pnpbp.2011.12.004.

Fowler, C. J., Jonsson, K.-O., Andersson, A., Juntunen, J., Järvinen, T., Vandevoorde, S., et al. (2003). Inhibition of C6 glioma cell proliferation by anandamide, 1-arachidonoylglycerol, and by a water soluble phosphate ester of anandamide: Variability in response and involvement of arachidonic acid. *Biochemical Pharmacology, 66*(5), 757–767.

Geldenhuys, W. J., & Van der Schyf, C. J. (2013). Rationally designed multi-targeted agents against neurodegenerative diseases. *Current Medicinal Chemistry, 20*(13), 1662–1672. https://doi.org/10.2174/09298673113209990112.

Gerdeman, G., & Lovinger, D. M. (2001). CB1 cannabinoid receptor inhibits synaptic release of glutamate in rat dorsolateral striatum. *Journal of Neurophysiology, 85*(1), 468–471.

Gerfen, C. R., & Surmeier, D. J. (2011). Modulation of striatal projection systems by dopamine. *Annual Review of Neuroscience, 34*(1), 441–466. https://doi.org/10.1146/annurev-neuro-061010-113641.

Glass, M., Dragunow, M., & Faull, R. L. M. (2000). The pattern of neurodegeneration in Huntington's disease: A comparative study of cannabinoid, dopamine, adenosine and GABA(A) receptor alterations in the human basal ganglia in Huntington's disease. *Neuroscience, 97*(3), 505–519. https://doi.org/10.1016/S0306-4522(00)00008-7.

Glass, M., Van Dellen, A., Blakemore, C., Hannan, A. J., & Faull, R. L. M. (2004). Delayed onset of Huntington's disease in mice in an enriched environment correlates with delayed loss of cannabinoid CB1 receptors. *Neuroscience, 123*(1), 207–212. https://doi.org/10.1016/S0306-4522(03)00595-5.

Gong, J. P., Onaivi, E. S., Ishiguro, H., Liu, Q. R., Tagliaferro, P. A., Brusco, A., et al. (2006). Cannabinoid CB2 receptors: Immunohistochemical localization in rat brain. *Brain Research, 1071*(1), 10–23. https://doi.org/10.1016/j.brainres.2005.11.035.

Gulyas, A. I., Cravatt, B. F., Bracey, M. H., Dinh, T. P., Piomelli, D., Boscia, F., et al. (2004). Segregation of two endocannabinoid-hydrolyzing enzymes into pre- and postsynaptic compartments in the rat hippocampus, cerebellum and amygdala. *European Journal of Neuroscience, 20*(2), 441–458. https://doi.org/10.1111/j.1460-9568.2004.03428.x.

Hampson, A. J., Grimaldi, M., Axelrod, J., & Wink, D. (1998). Cannabidiol and (−) 9-tetrahydrocannabinol are neuroprotective antioxidants. *Proceedings of the National Academy of Sciences of the United States of America, 95*(14), 8268–8273. https://doi.org/10.1073/pnas.95.14.8268.

Hansen, H. H., Azcoitia, I., Pons, S., Romero, J., García-Segura, L. M., Ramos, J. A., et al. (2002). Blockade of cannabinoid CB1receptor function protects against in vivo disseminating brain damage following NMDA-induced excitotoxicity. *Journal of Neurochemistry*, *82*(1), 154–158. https://doi.org/10.1046/j.1471-4159.2002.00961.x.

Hashimotodani, Y., Ohno-Shosaku, T., & Kano, M. (2007). Presynaptic Monoacylglycerol lipase activity determines basal endocannabinoid tone and terminates retrograde endocannabinoid signaling in the Hippocampus. *Journal of Neuroscience*, *27*(5), 1211–1219. https://doi.org/10.1523/JNEUROSCI.4159-06.2007.

Hedreen, J. C., Peyser, C. E., Folstein, S. E., & Ross, C. A. (1991). Neuronal loss in layers V and VI of cerebral cortex in Huntington's disease. *Neuroscience Letters*, *133*(2), 257–261. https://doi.org/10.1016/0304-3940(91)90583-F.

Heinsen, H., Strik, M., Bauer, M., Luther, K., Ulmar, G., Gangnus, D., et al. (1994). Cortical and striatal neurone number in Huntington's disease. *Acta Neuropathologica*, *88*(4), 320–333. https://doi.org/10.1007/BF00310376.

Hohmann, A. G., & Herkenham, M. (2000). Localization of cannabinoid CB1receptor mRNA in neuronal subpopulations of rat striatum: A double-label in situ hybridization study. *Synapse*, *37*(1), 71–80. https://doi.org/10.1002/(SICI)1098-2396(200007)37:1<71::AID-SYN8>3.0.CO;2-K.

Iuvone, T., Esposito, G., De Filippis, D., Scuderi, C., & Steardo, L. (2009). Cannabidiol: A promising drug for neurodegenerative disorders? *CNS Neuroscience and Therapeutics*, *15*(1), 65–75. https://doi.org/10.1111/j.1755-5949.2008.00065.x.

Javed, H., Azimullah, S., Haque, M. E., & Ojha, S. K. (2016). Cannabinoid type 2 (CB2) receptors activation protects against oxidative stress and neuroinflammation associated dopaminergic neurodegeneration in rotenone model of parkinson's disease. *Frontiers in Neuroscience*, *10*, 321. https://doi.org/10.3389/fnins.2016.00321.

Kallendrusch, S., Kremzow, S., Nowicki, M., Grabiec, U., Winkelmann, R., Benz, A., et al. (2013). The G protein-coupled receptor 55 ligand 1-α-lysophosphatidylinositol exerts microglia-dependent neuroprotection after excitotoxic lesion. *Glia*, *61*(11), 1822–1831. https://doi.org/10.1002/glia.22560.

Karanian, D. A., Karim, S. L., Wood, J. T., Williams, J. S., Lin, S., Makriyannis, A., et al. (2007). Endocannabinoid enhancement protects against Kainic acid-induced seizures and associated brain damage. *Journal of Pharmacology and Experimental Therapeutics*, *322*(3), 1059–1066. https://doi.org/10.1124/jpet.107.120147.

Katona, I., & Freund, T. F. (2008). Endocannabinoid signaling as a synaptic circuit breaker in neurological disease. *Nature Medicine*, *14*(9), 923–930. https://doi.org/10.1038/nm.f.1869.

Kaur, K., Gill, J. S., Bansal, P. K., & Deshmukh, R. (2017). Neuroinflammation—a major cause for striatal dopaminergic degeneration in Parkinson's disease. *Journal of the Neurological Sciences*, *381*, 308–314. https://doi.org/10.1016/j.jns.2017.08.3251.

Khaspekov, L. G., Verca, M. S. B., Frumkina, L. E., Hermann, H., Marsicano, G., & Lutz, B. (2004). Involvement of brain-derived neurotrophic factor in cannabinoid receptor-dependent protection against excitotoxicity. *European Journal of Neuroscience*, *19*(7), 1691–1698. https://doi.org/10.1111/j.1460-9568.2004.03285.x.

Kim, S. R., Lee, D. Y., Chung, E. S., Oh, U. T., Kim, S. U., & Jin, B. K. (2005). Transient receptor potential Vanilloid subtype 1 mediates cell death of mesencephalic dopaminergic neurons in vivo and in vitro. *Journal of Neuroscience*, *25*(3), 662–671. https://doi.org/10.1523/JNEUROSCI.4166-04.2005.

Kofalvi, A. (2005). Involvement of cannabinoid receptors in the regulation of neurotransmitter release in the rodent striatum: A combined immunochemical and pharmacological analysis. *Journal of Neuroscience*, *25*(11), 2874–2884. https://doi.org/10.1523/JNEUROSCI.4232-04.2005.

Lanciego, J. L., Barroso-Chinea, P., Rico, A. J., Conte-Perales, L., Callén, L., Roda, E., et al. (2011). Expression of the mRNA coding the cannabinoid receptor 2 in the pallidal complex of *Macaca fascicularis*. *Journal of Psychopharmacology*, *25*(1), 97–104. https://doi.org/10.1177/0269881110367732.

Laprairie, R. B., Warford, J. R., Hutchings, S., Robertson, G. S., Kelly, M. E. M., & Denovan-Wright, E. M. (2014). The cytokine and endocannabinoid systems are co-regulated by NF-κB p65/RelA in cell culture and transgenic mouse models of Huntington's disease and in striatal tissue from Huntington's disease patients. *Journal of Neuroimmunology*, *267*(1–2), 61–72. https://doi.org/10.1016/j.jneuroim.2013.12.008.

Lastres-Becker, I., Bizat, N., Boyer, F., Hantraye, P., Fernández-Ruiz, J., & Brouillet, E. (2004). Potential involvement of cannabinoid receptors in 3-nitropropionic acid toxicity in vivo. *Neuroreport*, *15*(15), 2375–2379. https://doi.org/10.1097/00001756-200410250-00015.

Lastres-Becker, I., Molina-Holgado, F., Ramos, J. A., Mechoulam, R., & Fernández-Ruiz, J. (2005). Cannabinoids provide neuroprotection against 6-hydroxydopamine toxicity in vivo and in vitro: Relevance to Parkinson's disease. *Neurobiology of Disease*, *19*(1–2), 96–107. https://doi.org/10.1016/j.nbd.2004.11.009.

Leweke, F. M., Piomelli, D., Pahlisch, F., Muhl, D., Gerth, C. W., Hoyer, C., et al. (2012). Cannabidiol enhances anandamide signaling and alleviates psychotic symptoms of schizophrenia. *Translational Psychiatry*, *2*(3), e94. https://doi.org/10.1038/tp.2012.15.

Liu, H., Song, Z., Liao, D., Zhang, T., Liu, F., Zhuang, K., et al. (2014). Neuroprotective effects of trans-Caryophyllene against Kainic acid induced seizure activity and oxidative stress in mice. *Neurochemical Research*, *40*(1), 118–123. https://doi.org/10.1007/s11064-014-1474-0.

López, A., Aparicio, N., Pazos, M. R., Grande, M. T., Barreda-Manso, M. A., Benito-Cuesta, I., et al. (2018). Cannabinoid CB2receptors in the mouse brain: Relevance for Alzheimer's disease. *Journal of Neuroinflammation*, *15*(1), 158. https://doi.org/10.1186/s12974-018-1174-9.

López-Sendón Moreno, J. L., García Caldentey, J., Trigo Cubillo, P., Ruiz Romero, C., García Ribas, G., Alonso Arias, M. A. A., et al. (2016). A double-blind, randomized, cross-over, placebo-controlled, pilot trial with Sativex in Huntington's disease. *Journal of Neurology*, *263*(7), 1390–1400. https://doi.org/10.1007/s00415-016-8145-9.

Lull, M. E., & Block, M. L. (2010). Microglial activation and chronic neurodegeneration. *Neurotherapeutics*, 7(4), 354–365. https://doi.org/10.1016/j.nurt.2010.05.014.

Ma, L., Jia, J., Niu, W., Jiang, T., Zhai, Q., Yang, L., et al. (2015). Mitochondrial CB1 receptor is involved in ACEA-induced protective effects on neurons and mitochondrial functions. *Scientific Reports*, *5*(1), 12440. https://doi.org/10.1038/srep12440.

MacDonald, M. E., Ambrose, C. M., Duyao, M. P., Myers, R. H., Lin, C., Srinidhi, L., et al. (1993). A novel gene containing a trinucleotide repeat that is expanded and unstable on Huntington's disease chromosomes. *Cell*, *72*(6), 971–983. https://doi.org/10.1016/0092-8674(93)90585-E.

Mailleux, P., & Vanderhaeghen, J. J. (1992). Distribution of neuronal cannabinoid receptor in the adult rat brain: A comparative receptor binding radioautography and in situ hybridization histochemistry. *Neuroscience*, *48*(3), 655–668. https://doi.org/10.1016/0306-4522(92)90409-U.

Malek, N., Popiolek-Barczyk, K., Mika, J., Przewlocka, B., & Starowicz, K. (2015). Anandamide, acting via CB2 receptors, alleviates LPS-induced neuroinflammation in rat primary microglial cultures. *Neural Plasticity*, *2015*, 130639. https://doi.org/10.1155/2015/130639.

Marsicano, G., Moosmann, B., Hermann, H., Lutz, B., & Behl, C. (2002). Neuroprotective properties of cannabinoids against oxidative stress: Role of the cannabinoid receptor CB1. *Journal of Neurochemistry*, *80*(3), 448–456. https://doi.org/10.1046/j.0022-3042.2001.00716.x.

Massano, J., & Bhatia, K. P. (2012). Clinical approach to Parkinson's disease: Features, diagnosis, and principles of management. *Cold Spring Harbor Perspectives in Medicine*, *2*(6), a008870. https://doi.org/10.1101/cshperspect.a008870.

Maya-López, M., Colín-González, A. L., Aguilera, G., de Lima, M. E., Colpo-Ceolin, A., Rangel-López, E., et al. (2017). Neuroprotective effect of WIN55,212-2 against 3-nitropropionic acid-induced toxicity in the rat brain: Involvement of CB1 and NMDA receptors. *American Journal of Translational Research*, *9*(2), 261–274. Retrieved from http://www.ncbi.nlm.nih.gov/pubmed/28337258.

Maya-López, M., Ruiz-Contreras, H. A., de Jesús Negrete-Ruíz, M., Martínez-Sánchez, J. E., Benítez-Valenzuela, J., Colín-González, A. L., et al. (2017). URB597 reduces biochemical, behavioral and morphological alterations in two neurotoxic models in rats. *Biomedicine and Pharmacotherapy*, *88*, 745–753. https://doi.org/10.1016/j.biopha.2017.01.116.

McCaw, E. A., Hu, H., Gomez, G. T., Hebb, A. L. O., Kelly, M. E. M., & Denovan-Wright, E. M. (2004). Structure, expression and regulation of the cannabinoid receptor gene (CB1) in Huntington's disease transgenic mice. *European Journal of Biochemistry*, *271*(23–24), 4909–4920. https://doi.org/10.1111/j.1432-1033.2004.04460.x.

Mechoulam, R., Panikashvili, D., & Shohami, E. (2002). Cannabinoids and brain injury: Therapeutic implications. *Trends in Molecular Medicine*, *8*(2), 58–61. https://doi.org/10.1016/S1471-4914(02)02276-1.

Mezey, E., Toth, Z. E., Cortright, D. N., Arzubi, M. K., Krause, J. E., Elde, R., et al. (2000). Distribution of mRNA for vanilloid receptor subtype 1 (VR1), and VR1-like immunoreactivity, in the central nervous system of the rat and human. *Proceedings of the National Academy of Sciences of the United States of America*, *97*(7), 3655–3660. https://doi.org/10.1073/pnas.97.7.3655.

Mievis, S., Blum, D., & Ledent, C. (2011). Worsening of Huntington disease phenotype in CB1 receptor knockout mice. *Neurobiology of Disease*, *42*(3), 524–529. https://doi.org/10.1016/j.nbd.2011.03.006.

Morera-Herreras, T., Miguelez, C., Aristieta, A., Ruiz-Ortega, J. Á., & Ugedo, L. (2012). Endocannabinoid modulation of dopaminergic motor circuits. *Frontiers in Pharmacology*, *3*, 110. https://doi.org/10.3389/fphar.2012.00110.

Muñoz-Arenas, G., Paz-Bermúdez, F., Báez-Cordero, A., Caballero-Florán, R., González-Hernández, B., Florán, B., et al. (2015). Cannabinoid CB1 receptors activation and coactivation with D2 receptors modulate GABAergic neurotransmission in the globus pallidus and increase motor asymmetry. *Synapse*, *69*(3), 103–114. https://doi.org/10.1002/syn.21796.

Munro, S., Thomas, K. L., & Abu-Shaar, M. (1993). Molecular characterization of a peripheral receptor for cannabinoids. *Nature*, *365*(6441), 61–65. https://doi.org/10.1038/365061a0.

Nadler, V., Mechoulam, R., & Sokolovsky, M. (1993). Blockade of45Ca2+influx through the N-methyl-d-aspartate receptor ion channel by the non-psychoactive cannabinoid HU-211. *Brain Research*, *622*(1–2), 79–85. https://doi.org/10.1016/0006-8993(93)90804-V.

Naidoo, V., Nikas, S. P., Karanian, D. A., Hwang, J., Zhao, J., Wood, J. A. T., et al. (2011). A new generation fatty acid amide hydrolase inhibitor protects against kainate-induced excitotoxicity. *Journal of Molecular Neuroscience*, *43*(3), 493–502. https://doi.org/10.1007/s12031-010-9472-4.

Navarrete, F., García-Gutiérrez, M. S., Aracil-Fernández, A., Lanciego, J. L., & Manzanares, J. (2018). Cannabinoid CB1 and CB2 receptors, and Monoacylglycerol lipase gene expression alterations in the basal ganglia of patients with Parkinson's disease. *Neurotherapeutics*, *15*(2), 1–11. https://doi.org/10.1007/s13311-018-0603-x.

Navarro, G., Morales, P., Rodríguez-Cueto, C., Fernández-Ruiz, J., Jagerovic, N., & Franco, R. (2016). Targeting cannabinoid CB2receptors in the central nervous system. Medicinal chemistry approaches with focus on neurodegenerative disorders. *Frontiers in Neuroscience*, *10*(Sep), 406. https://doi.org/10.3389/fnins.2016.00406.

Naydenov, A. V., Sepers, M. D., Swinney, K., Raymond, L. A., Palmiter, R. D., & Stella, N. (2014). Genetic rescue of CB1receptors on medium spiny neurons prevents loss of excitatory striatal synapses but not motor impairment in HD mice. *Neurobiology of Disease*, *71*, 140–150. https://doi.org/10.1016/j.nbd.2014.08.009.

Núñez, E., Benito, C., Tolón, R. M., Hillard, C. J., Griffin, W. S. T., & Romero, J. (2008). Glial expression of cannabinoid CB2 receptors and fatty acid amide hydrolase are beta amyloid-linked events in Down's syndrome. *Neuroscience*, *151*(1), 104–110. https://doi.org/10.1016/j.neuroscience.2007.10.029.

Nussbaum, R. L., & Ellis, C. E. (2003). Alzheimer's disease and Parkinson's disease. *New England Journal of Medicine*, *348*(14), 1356–1364. https://doi.org/10.1056/NEJM2003ra020003.

O'Sullivan, S. E. (2016). An update on PPAR activation by cannabinoids. *British Journal of Pharmacology*, *173*(12), 1899–1910. https://doi.org/10.1111/bph.13497.

Oh, Y. T., Lee, J. Y., Lee, J., Lee, J. H., Kim, J. E., Ha, J., et al. (2010). Oleamide suppresses lipopolysaccharide-induced expression of iNOS and COX-2 through inhibition of NF-κB activation in BV2 murine microglial cells. *Neuroscience Letters*, *474*(3), 148–153. https://doi.org/10.1016/j.neulet.2010.03.026.

Ojha, S., Javed, H., Azimullah, S., & Haque, M. E. (2016). β-Caryophyllene, a phytocannabinoid attenuates oxidative stress, neuroinflammation, glial activation, and salvages dopaminergic neurons in a rat model of Parkinson disease. *Molecular and Cellular Biochemistry*, *418*(1–2), 59–70. https://doi.org/10.1007/s11010-016-2733-y.

Onaivi, E. S., Ishiguro, H., Gong, J. P., Patel, S., Perchuk, A., Meozzi, P. A., et al. (2006). Discovery of the presence and functional expression of cannabinoid CB2 receptors in brain. *Annals of the New York Academy of Sciences*, *1074*(1), 514–536. https://doi.org/10.1196/annals.1369.052.

Ouchi, Y., Yoshikawa, E., Sekine, Y., Futatsubashi, M., Kanno, T., Ogusu, T., et al. (2005). Microglial activation and dopamine terminal loss in early Parkinson's disease. *Annals of Neurology*, *57*(2), 168–175. https://doi.org/10.1002/ana.20338.

Palazuelos, J., Aguado, T., Pazos, M. R., Julien, B., Carrasco, C., Resel, E., et al. (2009). Microglial CB2 cannabinoid receptors are neuroprotective in Huntington's disease excitotoxicity. *Brain*, *132*(11), 3152–3164. https://doi.org/10.1093/brain/awp239.

Pazos, M. R., Mohammed, N., Lafuente, H., Santos, M., Martínez-Pinilla, E., Moreno, E., et al. (2013). Mechanisms of cannabidiol neuroprotection in hypoxic-ischemic newborn pigs: Role of 5HT1Aand CB2 receptors. *Neuropharmacology*, *71*, 282–291. https://doi.org/10.1016/j.neuropharm.2013.03.027.

Pelição, R., Santos, M. C., Freitas-Lima, L. C., Meyrelles, S. S., Vasquez, E. C., Nakamura-Palacios, E. M., et al. (2016). URB597 inhibits oxidative stress induced by alcohol binging in the prefrontal cortex of adolescent rats. *Neuroscience Letters*, *624*, 17–22. https://doi.org/10.1016/j.neulet.2016.04.068.

Pintor, A., Tebano, M. T., Martire, A., Grieco, R., Galluzzo, M., Scattoni, M. L., et al. (2006). The cannabinoid receptor agonist WIN 55,212-2 attenuates the effects induced by quinolinic acid in the rat striatum. *Neuropharmacology*, *51*(5), 1004–1012. https://doi.org/10.1016/j.neuropharm.2006.06.013.

Piomelli, D. (2003). The molecular logic of endocannabinoid signalling. *Nature Reviews Neuroscience*, *4*(11), 873–884. https://doi.org/10.1038/nrn1247.

Poewe, W., & Mahlknecht, P. (2010). The clinical progression of Parkinson's disease. *Parkinsonism & Related Disorders*, *15*(Suppl. 4), S28–S32. https://doi.org/10.1016/S1353-8020(09)70831-4.

Price, D. A., Martinez, A. A., Seillier, A., Koek, W., Acosta, Y., Fernandez, E., et al. (2009). WIN55,212-2, a cannabinoid receptor agonist, protects against nigrostriatal cell loss in the 1-methyl-4-phenyl-1,2,3,6-tetrahydropyridine mouse model of Parkinson's disease. *European Journal of Neuroscience*, *29*(11), 2177–2186. https://doi.org/10.1111/j.1460-9568.2009.06764.x.

Ramirez, B. G. (2005). Prevention of Alzheimer's disease pathology by cannabinoids: Neuroprotection mediated by blockade of microglial activation. *Journal of Neuroscience*, *25*(8), 1904–1913. https://doi.org/10.1523/JNEUROSCI.4540-04.2005.

Ramsey, C. P., & Tansey, M. G. (2014). A survey from 2012 of evidence for the role of neuroinflammation in neurotoxin animal models of Parkinson's disease and potential molecular targets. *Experimental Neurology*, *256*, 126–132. https://doi.org/10.1016/j.expneurol.2013.05.014.

Rangel-López, E., Colín-González, A. L., Paz-Loyola, A. L., Pinzón, E., Torres, I., Serratos, I. N., et al. (2015). Cannabinoid receptor agonists reduce the short-term mitochondrial dysfunction and oxidative stress linked to excitotoxicity in the rat brain. *Neuroscience*, *285*, 97–106. https://doi.org/10.1016/j.neuroscience.2014.11.016.

Reiner, A., Albin, R. L., Anderson, K. D., D'Amato, C. J., Penney, J. B., & Young, A. B. (1988). Differential loss of striatal projection neurons in Huntington disease. *Proceedings of the National Academy of Sciences of the United States of America*, *85*(15), 5733–5737. https://doi.org/10.1073/pnas.85.15.5733.

Romero, J., Lastres-Becker, I., De Miguel, R., Berrendero, F., Ramos, J. A., & Fernández-Ruiz, J. (2002). The endogenous cannabinoid system and the basal ganglia: Biochemical, pharmacological, and therapeutic aspects. *Pharmacology and Therapeutics*, *95*(2), 137–152. https://doi.org/10.1016/S0163-7258(02)00253-X.

Saft, C., Von Hein, S. M., Lucke, T., Thiels, C., Peball, M., Djamshidian, A., et al. (2018). Cannabinoids for treatment of dystonia in Huntington's disease. *Journal of Huntington's Disease*, 7(2), 167–173. https://doi.org/10.3233/JHD-170283.

Sagredo, O., García-Arencibia, M., De Lago, E., Finetti, S., Decio, A., & Fernández-Ruiz, J. (2007). Cannabinoids and neuroprotection in basal ganglia disorders. *Molecular Neurobiology*, *36*(1), 82–91. https://doi.org/10.1007/s12035-007-0004-3.

Sagredo, O., González, S., Aroyo, I., Pazos, M. R., Benito, C., Lastres-Becker, I., et al. (2009). Cannabinoid CB2receptor agonists protect the striatum against malonate toxicity: Relevance for Huntington's disease. *Glia*, *57*(11), 1154–1167. https://doi.org/10.1002/glia.20838.

Sagredo, O., Pazos, M. R., Satta, V., Ramos, J. A., Pertwee, R. G., & Fernández-Ruiz, J. (2011). Neuroprotective effects of phytocannabinoid-based medicines in experimental models of Huntington's disease. *Journal of Neuroscience Research*, *89*(9), 1509–1518. https://doi.org/10.1002/jnr.22682.

Schapira, A. H. (2008). Mitochondria in the aetiology and pathogenesis of Parkinson's disease. *The Lancet Neurology*, 7(1), 97–109. https://doi.org/10.1016/S1474-4422(07)70327-7.

Schapira, A. H. V., Olanow, C. W., Greenamyre, J. T., & Bezard, E. (2014). Slowing of neurodegeneration in Parkinson's disease and Huntington's disease: Future therapeutic perspectives. *The Lancet*, *384*(9942), 545–555. https://doi.org/10.1016/S0140-6736(14)61010-2.

Scuderi, C., Steardo, L., & Esposito, G. (2014). Cannabidiol promotes amyloid precursor protein ubiquitination and reduction of beta amyloid expression in SHSY5YAPP+ cells through PPARγ involvement. *Phytotherapy Research*, *28*(7), 1007–1013. https://doi.org/10.1002/ptr.5095.

Serpa, A., Pinto, I., Bernardino, L., & Cascalheira, J. F. (2015). Combined neuroprotective action of adenosine A1and cannabinoid CB1receptors against NMDA-induced excitotoxicity in the hippocampus. *Neurochemistry International*, *87*, 106–109. https://doi.org/10.1016/j.neuint.2015.06.005.

Sheng, W. S., Hu, S., Ni, H. T., Rock, R. B., & Peterson, P. K. (2009). WIN55,212-2 inhibits production of CX3CL1 by human astrocytes: Involvement of p38 MAP kinase. *Journal of Neuroimmune Pharmacology*, *4*(2), 244–248. https://doi.org/10.1007/s11481-009-9147-5.

Sierra, S., Luquin, N., Rico, A. J., Gómez-Bautista, V., Roda, E., Dopeso-Reyes, I. G., et al. (2015). Detection of cannabinoid receptors CB1 and CB2 within basal ganglia output neurons in macaques: Changes following experimental parkinsonism. *Brain Structure and Function*, *220*(5), 2721–2738. https://doi.org/10.1007/s00429-014-0823-8.

Smith, S. R., Terminelli, C., & Denhardt, G. (2000). Effects of cannabinoid receptor agonist and antagonist ligands on production of inflammatory cytokines and anti-inflammatory interleukin-10 in endotoxemic mice. *The Journal of Pharmacology and Experimental Therapeutics*, *293*(1), 136–150. Retrieved from http://www.ncbi.nlm.nih.gov/pubmed/10734163.

Solinas, M., Justinova, Z., Goldberg, S. R., & Tanda, G. (2006). Anandamide administration alone and after inhibition of fatty acid amide hydrolase (FAAH) increases dopamine levels in the nucleus accumbens shell in rats. *Journal of Neurochemistry*, *98*(2), 408–419. https://doi.org/10.1111/j.1471-4159.2006.03880.x.

Starowicz, K., Nigam, S., & Di Marzo, V. (2007). Biochemistry and pharmacology of endovanilloids. *Pharmacology and Therapeutics*, *114*(1), 13–33. https://doi.org/10.1016/j.pharmthera.2007.01.005.

Sugiura, T., Kishimoto, S., Oka, S., & Gokoh, M. (2006). Biochemistry, pharmacology and physiology of 2-arachidonoylglycerol, an endogenous cannabinoid receptor ligand. *Progress in Lipid Research*, *45*(5), 405–446. https://doi.org/10.1016/j.plipres.2006.03.003.

Sun, S., Hu, F., Wu, J., & Zhang, S. (2017). Cannabidiol attenuates OGD/R-induced damage by enhancing mitochondrial bioenergetics and modulating glucose metabolism via pentose-phosphate pathway in hippocampal neurons. *Redox Biology*, *11*, 577–585. https://doi.org/10.1016/j.redox.2016.12.029.

Takahashi, K. A., & Castillo, P. E. (2006). The CB1 cannabinoid receptor mediates glutamatergic synaptic suppression in the hippocampus. *Neuroscience*, *139*(3), 795–802. https://doi.org/10.1016/j.neuroscience.2006.01.024.

Tanimura, A., Yamazaki, M., Hashimotodani, Y., Uchigashima, M., Kawata, S., Abe, M., et al. (2010). The endocannabinoid 2-Arachidonoylglycerol produced by diacylglycerol lipase α mediates retrograde suppression of synaptic transmission. *Neuron*, *65*(3), 320–327. https://doi.org/10.1016/j.neuron.2010.01.021.

Teismann, P., & Schulz, J. B. (2004). Cellular pathology of Parkinson's disease: Astrocytes, microglia and inflammation. *Cell and Tissue Research*, *318*(1), 149–161. https://doi.org/10.1007/s00441-004-0944-0.

Thomas, B. (2009). Parkinson's disease: From molecular pathways in disease to therapeutic approaches. *Antioxidants & Redox Signaling*, *11*(9), 2077–2082. https://doi.org/10.1089/ars.2009.2697.

van der Stelt, M., & Di Marzo, V. (2003). The endocannabinoid system in the basal ganglia and in the mesolimbic reward system: Implications for neurological and psychiatric disorders. *European Journal of Pharmacology*, *480*(1–3), 133–150. https://doi.org/10.1016/j.ejphar.2003.08.101.

van der Stelt, M., Veldhuis, W. B., Bär, P. R., Veldink, G. A., Vliegenthart, J. F. G., & Nicolay, K. (2001). Neuroprotection by Δ^9-tetrahydrocannabinol, the main active compound in marijuana, against Ouabain-induced in vivo excitotoxicity. *The Journal of Neuroscience*, *21*(17), 6475–6479. https://doi.org/10.1523/JNEUROSCI.21-17-06475.2001.

van der Stelt, M., Veldhuis, W. B., Maccarrone, M., Bär, P. R., Nicolay, K., Veldink, G. A., et al. (2002). Acute neuronal injury, excitotoxicity, and the endocannabinoid system. *Molecular Neurobiology*, *26*(2–3), 317–346. https://doi.org/10.1385/MN:26:2-3:317.

Van Laere, K., Casteels, C., Dhollander, I., Goffin, K., Grachev, I., Bormans, G., et al. (2010). Widespread decrease of type 1 cannabinoid receptor availability in Huntington disease in vivo. *Journal of Nuclear Medicine: Official Publication, Society of Nuclear Medicine*, *51*(9), 1413–1417. https://doi.org/10.2967/jnumed.110.077156.

Van Sickle, M. D., Duncan, M., Kingsley, P. J., Mouihate, A., Urbani, P., Mackie, K., et al. (2005). Neuroscience: Identification and functional characterization of brainstem cannabinoid CB2 receptors. *Science*, *310*(5746), 329–332. https://doi.org/10.1126/science.1115740.

Varvel, S. A., Cravatt, B. F., Engram, A. E., & Lichtman, A. H. (2006). Fatty acid amide hydrolase (−/−) mice exhibit an increased sensitivity to the disruptive effects of anandamide or oleamide in a working memory water maze task. *The Journal of Pharmacology and Experimental Therapeutics*, *317*(1), 251–257. https://doi.org/10.1124/jpet.105.095059.

Veldhuis, W. B., van der Stelt, M., Wadman, M. W., van Zadelhoff, G., Maccarrone, M., Fezza, F., et al. (2003). Neuroprotection by the endogenous cannabinoid anandamide and arvanil against *in vivo* excitotoxicity in the rat: Role of vanilloid receptors and lipoxygenases. *The Journal of Neuroscience*, *23*(10), 4127–4133. https://doi.org/10.1523/JNEUROSCI.23-10-04127.2003.

Vrechi, T. A., Crunfli, F., Costa, A. P., & Torrão, A. S. (2018). Cannabinoid receptor type 1 agonist ACEA protects neurons from death and attenuates endoplasmic reticulum stress-related apoptotic pathway signaling. *Neurotoxicity Research*, *33*(4), 846–855. https://doi.org/10.1007/s12640-017-9839-1.

Walter, L., Franklin, A., Witting, A., Wade, C., Xie, Y., Kunos, G., et al. (2003). Nonpsychotropic cannabinoid receptors regulate microglial cell migration. *The Journal of Neuroscience*, *23*(4), 1398–1405. https://doi.org/10.1523/JNEUROSCI.23-04-01398.2003.

Yang, L., Rozenfeld, R., Wu, D., Devi, L. A., Zhang, Z., & Cederbaum, A. (2014). Cannabidiol protects liver from binge alcohol-induced steatosis by mechanisms including inhibition of oxidative stress and increase in autophagy. *Free Radical Biology and Medicine*, *68*, 260–267. https://doi.org/10.1016/j.freeradbiomed.2013.12.026.

Zeissler, M.-L., Eastwood, J., McCorry, K., Hanemann, O. C., Zajicek, J. P., & Carroll, C. B. (2016). Delta-9-tetrahydrocannabinol protects against MPP+ toxicity in SH-SY5Y cells by restoring proteins involved in mitochondrial biogenesis. *Oncotarget*, 7(29), 46603–46614. https://doi.org/10.18632/oncotarget.10314.

Zhang, H.-Y., Gao, M., Liu, Q.-R., Bi, G.-H., Li, X., Yang, H.-J., et al. (2014). Cannabinoid CB 2 receptors modulate midbrain dopamine neuronal activity and dopamine-related behavior in mice. *Proceedings of the National Academy of Sciences*, *111*(46), E5007–E5015. https://doi.org/10.1073/pnas.1413210111.

CHAPTER SIX

Dysfunction of serotonergic neurons in Parkinson's disease and dyskinesia

Sergio Vegas-Suarez[a,b], Elena Paredes-Rodriguez[a,b], Asier Aristieta[c,d], Jose V. Lafuente[e,f], Cristina Miguelez[a,b], Luisa Ugedo[a,b,]*

[a]Department of Pharmacology, Faculty of Medicine and Nursing, University of the Basque Country (UPV/EHU), Leioa, Spain
[b]Neurodegenerative Diseases, Biocruces Health Research Institute, Barakaldo, Spain
[c]Université de Bordeaux, Institut des Maladies Neurodégénératives, UMR 5293, Bordeaux, France
[d]Centre National de la Recherche Scientifique, Institut des Maladies Neurodégénératives, UMR 5293, Bordeaux, France
[e]Department of Neurosciences, Faculty of Medicine and Nursing, University of the Basque Country (UPV/EHU), Leioa, Spain
[f]Nanosurgery, Biocruces Health Research Institute, Barakaldo, Spain
*Corresponding author: e-mail address: luisa.ugedo@ehu.eus

Contents

Abstract

Parkinson's disease (PD) is characterized by the degeneration of dopaminergic neurons in the substantia nigra, the depletion of striatal dopamine and the presence of Lewy aggregates containing alpha-synuclein. Clinically, there are motor impairments involving cardinal movement symptoms, bradykinesia, resting tremor, muscle rigidity, and postural abnormalities, along with non-motor symptoms such as sleep, behavior and mood disorders. The current treatment for PD focuses on restoring dopaminergic neurotransmission by L-3,4-dihydroxyphenylalanine (levodopa), which loses therapeutic efficacy and induces disabling abnormal involuntary movements known as levodopa-induced dyskinesia (LID) after several years. Evidence indicates that the pathophysiology of both PD and LID disorders is also associated with the dysfunctional activity of the serotonergic (5-HT) neurons that may be responsible for motor and non-motor

International Review of Neurobiology, Volume 146
ISSN 0074-7742
https://doi.org/10.1016/bs.irn.2019.06.013

disturbances. The main population of 5-HT neurons is located in the dorsal raphe nuclei (DRN), which provides extensive innervation to almost the entire neuroaxis and controls multiple functions in the brain. The degeneration of DRN 5-HT neurons occurs in early PD. These neurons can also take exogenous levodopa to transform it into dopamine, which may disturb neuron activity. This review will provide an overview of the underlying mechanisms responsible for 5-HT dysfunction and its clinical relevance in PD and dyskinesia.

1. Introduction

Parkinson's disease (PD), the second most common neurodegenerative disease, is a chronic neurological disorder with a high worldwide incidence level, which is growing rapidly due to the aging of the world's population. In 2005, the estimated number of people with PD worldwide was 4.9 million (20/100,000), and this number will double by 2030 (Dorsey & Bloem, 2018). PD is characterized by a progressive degeneration of dopamine (DA) neurons located in the substantia nigra pars compacta, causing a massive loss of DA in the striatum (Kish, Shannak, & Hornykiewicz, 1988) and the widespread presence of intracellular Lewy bodies and Lewy neurites containing aggregates of α-synuclein in many brain regions. Clinically, PD is characterized by motor features including bradykinesia, resting tremor, muscular rigidity and postural and gait impairment, which are considered the cardinal symptoms; these symptoms first manifest when the loss of DA-producing neurons is approximately 60–80%. However, the clinical spectrum of PD is far greater than those motor symptoms since non-motor features, including olfactory dysfunction, cognitive impairment, psychiatric symptoms, sleep disorders, autonomic dysfunction, pain and fatigue, are also present and may precede the classic motor features by years (Kalia & Lang, 2015). Consequently, it is believed that not only the DA neurons from the substantia nigra but also other regions and neurotransmitters, such as the noradrenergic and the serotonergic (5-HT) systems, are affected in PD.

The most common treatments in clinical practice focus on DA replacement with the DA precursor levodopa, DA metabolism inhibitors or DA receptor agonists that reduce the majority of motor symptoms, but not all. Specifically, neither tremor nor most of the non-motor features improve with DA treatment. Additionally, the long-term administration of 3,4-dihydroxyphenylalanine (levodopa) induces abnormal involuntary movements known as levodopa-induced dyskinesia (LID), which are potentially disabling and affect up to 40% of PD patients within 5 years of treatment (Ahlskog & Muenter, 2001). The underlying mechanisms of LID are poorly understood, but there is vast consensus that they result from dysregulated DA neurotransmission (Huot, Johnston, Koprich, Fox, & Brotchie, 2013). In this sense, the 5-HT system may play a relevant role since levodopa is transformed into DA, stored and released by the 5-HT terminals (for review, see Miguelez, Benazzouz, Ugedo, & De Deurwaerdère, 2017).

PD is now viewed as a multisystem disorder, involving several neurotransmitters and multiple neuroanatomical areas. Indeed, experimental and clinical findings have provided compelling evidence for the prominent role of the 5-HT system in the pathophysiology of PD. Conversely, numerous studies have shown 5-HT degeneration in PD, which we will summarize in this review.

2. Why is the 5-HT system important in PD?

The 5-HT system is widely distributed through the brain and fulfills a significant role in the modulation of multiple central nervous system functions such as mood, emotion, sleep, appetite and motor activity. The system is also involved in the etiology of numerous neuropsychiatric disorders such as depression and anxiety. 5-HT is synthesized from tryptophan by the rate-limiting enzyme tryptophan hydroxylase and accumulates in secretor granules by a vesicular monoamine transporter before being released to the synaptic cleft. In the central nervous system, the action of 5-HT is terminated either via neuronal uptake by the 5-HT reuptake transporter (SERT) located in 5-HT axon terminals or by the enzymatic metabolism transformation (MAO) to 5-hydroxyindoleacetaldehyde and 5-hydroxyindoleacetic acid. 5-HT acts via 5-HT receptors, which are classified into seven families and at least 14 different subtypes, including G-protein-coupled receptor subtypes (5-HT_1, 5-HT_2, 5-HT_{4-7}) and ligand-gated ion channels (5-HT_3). The main assemblages of 5-HT-synthesizing neurons are located in the brainstem raphe nuclei whose neurons receive efferents from the peripheral and the central nervous system, including the substantia nigra. Neurons from the DRN send descending and ascending projections to innervate almost all brain areas, including the basal ganglia nuclei (BG). The BG form a complex, highly organized subcortical network that connects the thalamus with the cortex and plays an important role in the control of motor behavior, emotion and cognition. It is believed that functional alterations in BG are relevant in the etiopathology of movement disorders such as PD. BG are composed of the striatum (caudal and putamen), the subthalamic nucleus, the internal and external segments of the globus pallidus and the substantia nigra, pars reticulata and pars compacta. This latter region provides DA input into the BG, and when it degenerates, it leads to neuronal dysfunction in these nuclei. In the classical BG model (Alexander, DeLong, & Strick, 1986), the loss of DA neurons in PD leads to disinhibition of the indirect pathway and a reduction in the inhibitory action of the direct pathway, resulting in elevated activity of the output nuclei (the substantia nigra, pars reticulata and the internal segment of the globus pallidus). The augmented activity of the BG output structures, which are inhibitory, reduces motor activity, resulting in PD motor symptomatology. By contrast, levodopa administered to treat PD symptoms causes the opposite effect of DA loss, producing a decrease in the BG output nuclei activity that might underlie LID symptoms (Aristieta, Ruiz-Ortega, Miguelez, Morera-Herreras, & Ugedo, 2016; Obeso et al., 2008).

The BG contain 5-HT, SERT and almost all 5-HT receptor subtypes that are differently distributed. However, the effect of 5-HT drugs on the BG is complex since they are the result of the direct activation of 5-HT receptors in the different nuclei plus the activation of those located on 5-HT neurons that project to the BG (for a review, see Miguelez, Morera-Herreras, Torrecilla, Ruiz-Ortega, & Ugedo, 2014). For example, when we investigated the effect of 5-HT drugs on subthalamic nucleus activity, we observed that the direct application of a 5-HT_{1A} or 5-HT_{2C} receptor agonist caused a marked inhibition or stimulation, respectively, while systemic administration did not change the neuron activity in anesthetized rats (Aristieta et al., 2014). Furthermore, when we applied buspirone, a partial 5-HT_{1A} agonist, it did not have an effect on parkinsonian rats, whereas it inhibited subthalamic neurons in control rats (Sagarduy et al., 2016).

Despite this complexity, it is well accepted that the 5-HT system modulates the activity of the BG, and therefore, it is considered an important target in the development of new strategies for treating motor symptoms in PD, mainly those associated with long-term levodopa administration. On the other hand, selective 5-HT reuptake inhibitors, when chronically used to treat depression and anxiety, can induce motor side effects (tardive dyskinesia, Parkinsonism, akathisia and dystonia) (Bilen, Saka, Ak, & Öztekin, 2008; Leo, 1996). Studies performed in neonatal rats have also shown that fluoxetine treatment reduces locomotor activity (Lee & Lee, 2012). Chronic fluoxetine treatment did not impair motor coordination and balance in rats, although it induced catalepsy (Aristieta et al., 2014). Other authors have reported that acute fluoxetine administration potentiates haloperidol-induced catalepsy (Tatara et al., 2012). The mechanism underlying these motor side effects may involve the functional interaction between the 5-HT system and the BG.

3. What causes toxicity in 5-HT neurons?

Although the mechanism responsible for neuronal loss in PD is not completely understood, it is assumed that Lewy pathology is involved in the dysfunction or neurodegeneration that takes place in the central and peripheral nervous system. The presence of Lewy inclusions at the *post-mortem* pathological examination is the gold standard for diagnosis of PD. According to Braak's hypothesis, in PD, Lewy bodies and Lewy neurites are found all along the brain, following a progression of six stages beginning from the periphery (stage 1) and following within the brain in a caudal-to rostral direction (stages 2 to 6). In stage 2, Lewy pathology is present in the forebrain, including the locus coeruleus, reticular formation, raphe nuclei and spinal cord, and in stage 3, it affects the midbrain where the substantia nigra is localized. *Post-mortem* studies from patients with PD during the early stages of Lewy pathology have found the intracellular accumulation of alpha-synuclein in the raphe nuclei (Braak, Rüb, Gai, & Del Tredici, 2003), along with a loss of 5-HT neurons (Halliday, Blumbergs, Cotton, Blessing, & Geffen, 1990; Kovacs et al., 2003) and dystrophic degeneration of 5-HT axons in the cortex, hippocampus and substantia nigra (Azmitia & Nixon, 2008). Alterations of several components of 5-HT neurotransmission, such as a reduction in 5-HT and its metabolite content, tryptophan hydroxylase and SERT have also been described in the striatum (Kish et al., 2008) and hypothalamus (Shannak et al., 1994).

Lewy pathology consists of abnormal folded protein cumulates inside neuronal bodies (Lewy bodies) and thread-like proteinaceous deposits within neurites (Lewy neurites). Both Lewy bodies and Lewy neurites contain numerous proteins with alpha-synuclein as the major component (Spillantini et al., 1997). Alpha-synuclein is a 140-amino acid soluble protein encoded by the SNCA gene. This protein is mainly found in the cytosol and is essential for neuronal development and function. In physiological conditions, alpha-synuclein intervenes in intracellular trafficking, membrane interaction and synaptic activity (Hernandez, Reed, & Singleton, 2016) and may regulate monoamine homeostasis in the brain (Wersinger & Sidhu, 2009). However, in PD and other Lewy pathologies, the overproduction and accumulation of this protein, which becomes insoluble and filamentous, may result in neurotoxicity. One of the mechanisms underlying neurotoxicity is the fact that when upregulated, alpha-synuclein can target other cellular components such as mitochondria (Bendor, Logan, & Edwards, 2013) and can cause neuronal damage by disrupting the membrane, protein degradation or mitochondrial function (for review, see Sulzer, 2010).

Regarding the 5-HT system, alpha-synuclein is expressed in 5-HT raphe nuclei neurons where, in physiological conditions, it decreases the activity of the 5-HT transporter by its sequestration from the cellular membrane (Wersinger, Rusnak, & Sidhu, 2006). In fact, alpha-synuclein knockdown in monoaminergic nuclei is followed by increased 5-HT release in several brain areas of mice (Alarcón-Arís et al., 2018). *In vitro* studies performed in PC12 cells over-expressing alpha-synuclein have suggested that neuronal 5-HT metabolites, such as 5-hydroxyindolacetaldehyde, induce the oligomerization of alpha-synuclein prior to aggregation (Jinsmaa, Cooney, Sullivan, Sharabi, & Goldstein, 2015). As in the DA system, the expression of alpha-synuclein aberrant inclusions may result in 5-HT dysfunction and degeneration. In this sense, in the experiments conducted by Wan et al. (2016), where viral mediated alpha-synuclein overexpression was induced in 5-HT neurons, there was a progressive degeneration in 5-HT axons and dendrites. Similarly, a BAC-alpha-synuclein transgenic rat model that developed widespread and progressive alpha-synuclein aggregation showed reduced levels of 5-HT and 5-HT_{1B} receptors together with the loss of 5-HT terminals in the hippocampus (Kohl et al., 2016).

In addition to Lewy pathology, β-amyloid peptide plaque burden is also found in the brains of patients with PD, which is mainly related to cognitive impairment (for review, see Petrou et al., 2015). These β-amyloid plaques may also contribute to the dysfunction or degeneration of the 5-HT system since they are composed of amyloid β peptide, which can form oligomers and induce cytotoxicity by interfering with cellular physiology (Iadanza, Jackson, Hewitt, Ranson, & Radford, 2018). Positron emission tomography (PET) scanning of SERT and β-amyloid has revealed a relationship between reduced 5-HT terminals in the cortex and the increased severity of cerebral β-amyloid plaque burden in PD patients (Kotagal, Spino, Bohnen, Koeppe, & Albin, 2018). Furthermore, patients who received treatment with drugs that increased 5-HT levels, such as 5-HT reuptake inhibitors, for at least 6 months showed lower levels of β-amyloid in cerebrospinal fluid and less risk of cognitive decline than those without 5-HT medication (Kotagal et al., 2018).

Dopamine loss may also be implicated in the alterations of the 5-HT system during the development of PD since the raphe nuclei are under the control of DA and receive afferents from the substantia nigra. Therefore, the 5-HT system must adapt to the lack of DA caused by the loss of DA neurons. This issue has been studied in animal models of PD using the local injection of the toxin 6-hydroxydopamine (6-OHDA) along the nigrostriatal pathway or the systemic injection of 1-methyl-4-phenyl-1,2,3,6-tetrahydropyridine (MPTP), which causes DA neuron death and reduction in DA levels and fibers in the striatum. Although results from these studies have produced heterogeneous results showing no changes, increased activity of 5-HT neurons or up- and downregulation of one 5-HT-receptor subtype, together, they show that the 5-HT system is affected by DA loss through complex mechanisms (for a review, see Miguelez et al., 2014; Gagnon et al., 2018). Interestingly, in transgenic mice expressing human mutant LRRK2 G2019S, a model of the most common genetic cause of PD, along with a progressive loss of DA fibers in the striatum, the 5-HT levels declined, and 5-HT_{1A} receptor expression increased in the hippocampus (Lim et al., 2018). Findings of stereological studies using MPTP parkinsonian monkeys revealed a sprouting of 5-HT axon varicosities in the striatum and globus pallidus, while the density and morphology of 5-HT neurons of the DRN were unchanged (Gagnon, Gregoire, Di Paolo, & Parent, 2016; Gagnon et al., 2018). With these results, the authors suggest the rearrangement of the surviving 5-HT neurons and sprouting of 5-HT axons (Gagnon et al., 2016, 2018).

Another important cause of 5-HT system damage in PD is long-term levodopa administration since it is taken up and transformed into DA by 5-HT neurons (for review, see Miguelez et al., 2017). In 5-HT neurons, levodopa-derived DA displaces 5-HT in exocytotic vesicles, is trapped in the neuron and may be degraded by the enzyme monoamine oxidase and auto-oxidation mechanisms, resulting in the production of reactive oxygen species (Stansley & Yamamoto, 2014) and the subsequent detrimental effects (Navailles, Bioulac, Gross, & De Deurwaerdère, 2011; Stansley & Yamamoto, 2015a). Thus, levodopa can promote oxidative stress, generate free radicals or toxic compounds (Stansley & Yamamoto, 2015b) and cause 5-HT cell death and the depletion of 5-HT content in a subregion of the DRN and the frontal cortex (Stansley & Yamamoto, 2014).

In summary, as it is summarized in Fig. 1, there are several factors associated with PD and chronic levodopa administration that produce changes, damage and dysfunction of the 5-HT system and that may be clinically relevant not only regarding motor and non-motor symptoms but also in terms of PD treatment.

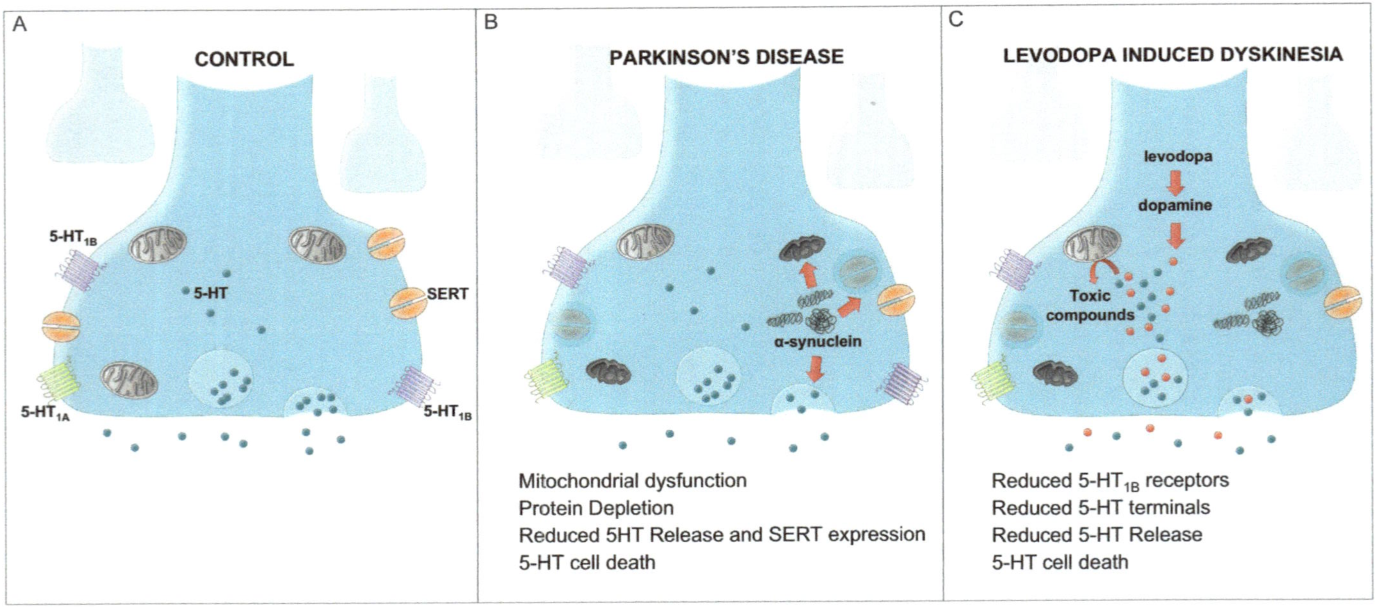

Fig. 1 Serotonergic neuron modificacions in Parkinson's disease and Levodopa-induced dyskinesia. (A) Control situation, (B) In Parkinson's disease, alpha-synuclein accumulates in 5-HT neurons and leads to neuron cell death by impacting mitochondrial function, protein depletion, 5-HT release and SERT expression. (C) During the chronic treatment with levodopa, apart from the modifications produced by the alpha-synuclein, levodopa-derived dopamine competes with 5-HT for entering in exocytosis vesicles and the accumulation of dopamine and 5-HT in the cytoplasm induces the synthesis of toxic compounds harmful for the cell.

4. Is the 5-HT system involved in the progression and symptoms of PD?

As mentioned previously, *post-mortem* studies from PD patients indicate an early 5-HT alteration in PD that has been corroborated *in vivo* using mainly imaging techniques. As occurs in animal research, the results are not homogeneous and vary depending on the brain region studied or the severity of the PD patients included. The availability of SERT was investigated using single-PET in nine early-stage patients and nine controls, resulting in no differences between the groups (Beucke et al., 2011). In contrast, a PET scanning study from 30 PD patients divided into groups according to the duration of the disease showed a reduction of SERT binding in striatal, brainstem, thalamus, hypothalamus and cortical regions in early PD stages that was greater in advanced PD but did not depend on DA therapy exposure (Politis et al., 2010). Furthermore, a meta-analysis of results obtained from 19 PET scanning studies in PD patients demonstrated that SERT binding was reduced in PD patients in the rostral and the caudal raphe, putamen, caudate, ventral striatum, thalamus, and hypothalamus. Additionally, the reduction of SERT binding in the putamen was associated with a longer duration of the disease (Pagano, Yousaf, & Politis, 2017).

Regarding motor symptoms and 5-HT function, neuroimaging studies in PD patients have also shown a reduction in the midbrain raphe 5-HT$_{1A}$ binding sites and a correlation between 5-HT$_{1A}$ density in the raphe and tremor but not rigidity or bradykinesia scores (Doder et al., 2003). Furthermore, SERT density in the raphe nuclei, thalamus and striatum correlated with the severity of resting tremor, which was lower in PD patients with tremor than in those without tremor (Caretti et al., 2008; Loane et al., 2013; Qamhawi et al., 2015).

In vivo clinical studies have also revealed a relationship between non-motor symptoms and the dysfunction of the 5-HT system. In fact, PD patients receiving DA-rich grafts showed improvement in motor symptoms, but they still suffered from non-motor symptoms. In these patients, functional imaging showed that the DA innervations were restored, but the 5-HT uptake binding was markedly reduced (Politis et al., 2012). In another study, plasma levels of 5-HT and 5-hydroxyindoleacetic acid were reduced, and their values correlated with depression and pain severity in PD patients (Tong et al., 2015). Data obtained from PD patients with depression have shown a correlation between the severity of the depressive symptoms and several 5-HT biomarkers, as 5-HT binding sites in raphe and limbic structures (Politis et al., 2010), 5-HT plasma levels (Tong et al., 2015) and loss of presynaptic SERT binding sites in the cortex (Maillet et al., 2016). There is also a correlation between sleep disturbance scores and SERT density in the dorsal (caudate and putamen) and ventral striatum, thalamus, hypothalamus and raphe nuclei (Wilson, Giordano, Turkheimer, Chaudhuri, & Politis, 2018). Anxiety severity scores showed a negative association with SERT binding sites in the thalamus (Joling, van den Heuvel, Berendse, Booij, & Vriend, 2018). The severity of apathy, one of the most common neuropsychiatric symptoms in PD, was related to specific 5-HT lesions within the right-side anterior caudate nucleus and the orbitofrontal cortex, as measured by PET technology of SERT in 15 apathetic and 15 non-apathetic drug-naïve *de novo* patients (Maillet et al., 2016). SERT binding density in limbic structures is also reduced in PD patients with fatigue compared to those without fatigue (Pavese, Metta, Bose, Chaudhuri, & Brooks, 2010), and 5-HT levels in the cerebrospinal fluid are inversely correlated with fatigue (Zuo et al., 2016), weight loss (Politis, Loane, Wu, Brooks, & Piccini, 2011) and visual hallucinations (Ballanger et al., 2010). Another important problem in PD is the high prevalence of sleep disturbances. Those patients with higher severity of sleep problems have lower SERT binding in the caudate, putamen, hypothalamus and raphe nuclei (Wilson et al., 2018).

A clinical trial applying 5-HT drugs also supports a relevant role for the 5-HT system in the clinical manifestations of PD. Citalopram, a SERT inhibitor, improved behavioral impulsivity only in PD patients with more advanced disease, but not in controls, which suggests a relationship between loss of 5-HT and the development of behavior impulsivity (Ye et al., 2014). In 2016, pimavanserin, a 5-HT_{2A}/5-HT_{2C} inverse agonist, was approved for the management of psychotic symptoms after showing efficacy and tolerability in two randomized controlled trials (Cummings et al., 2014; Meltzer et al., 2010). In an open-label multicenter randomized study, treatment with SERT improved depression and gait instability in PD patients (Takahashi et al., 2019). Low doses of clozapine, which acts as a 5-HT_{1A} and 5-HT_{2A}/5-HT_{2C} antagonist, improved tremor in PD patients, suggesting that these receptors are involved in the beneficial effect of the drug (see review by Fox, 2013). On the other hand, several clinical reports have highlighted that chronic treatment with selective 5-HT reuptake inhibitors, such as fluoxetine, can induce extrapyramidal side effects (*e.g.*, tardive dyskinesia or Parkinsonism), akathisia and/or dystonia (Bilen et al., 2008; Leo, 1996). Findings from preclinical investigations also show that 5-HT drugs have some beneficial effect on the motor and non-motor symptoms of Parkinsonism. The activation of 5-HT_{1A} receptors reduced catalepsy, bradykinesia and hypo-locomotion in 6-OHDA lesioned rats (for review, see Huang et al., 2017). In addition, 5-HT_{1A} activation reduced levodopa-induced rotational behavior (Inden et al., 2012), and treatment with fluvoxamine, a 5-HT uptake inhibitor, relieved cognitive impairment in a rat model of PD (Dallé, Daniels, & Mabandla, 2017).

In summary, most studies indicate that in PD, there is a reduction in the activity of the 5-HT system that depends on the progression of the illness and is relevant to motor and non-motor clinical manifestations. On the other hand, clinical and preclinical studies support the efficacy of 5-HT-based therapies for treating some motor and non-motor symptoms but also highlight the worsening of motor performance produced by these drugs.

5. Why is the 5-HT system important in levodopa-induced dyskinesia?

In the last 50 years, pharmacological replacement with levodopa has been the gold-standard treatment for PD. This treatment provides the greatest symptomatic benefit, but its long-term use is associated with motor complications (dyskinesias and motor fluctuations) and decreased efficacy. The appearance of these abnormal involuntary movements complicates the pharmacological treatment and highly affects patients' quality of life. As mentioned previously, when levodopa is used to treat PD, as there are few DA terminals, it is taken up and metabolized to DA in 5-HT terminals. This effect may be dangerous (toxic) to the 5-HT system, leading to adaptive changes that may contribute to the development of dyskinesias.

In this regard, although a wide range of publications support that levodopa induces the sprouting of 5-HT fibers in the brain and relates it with motor control, other authors also found the opposite results (see reviewed by Miguelez et al., 2014, 2017). For example, one study performed in two groups of patients, dyskinetic and non-dyskinetic, showed higher 5-HT hyper-innervation in the external globus pallidus and caudate only in the dyskinetic group, which directly correlated with the severity of motor complications (Rylander et al., 2010). In contrast, other studies have shown that the striatal *post-mortem* content of 5-HT and SERT did not differ significantly between dyskinetic and non-dyskinetic cases (Calon et al., 2003; Kish et al., 2008), and chronic levodopa treatment did not influence SERT expression (Politis et al., 2010). Additionally, some studies showed no changes in 5-HT_{1A} receptor levels in different brain areas, including the BG (Huot et al., 2012). In a recent systematic review of PET studies performed in PD patients, the authors concluded that dyskinetic PD patients have an increased 5-HT/DA transporter ratio, suggesting that when DA innervation in the striatum is low, the 5-HT system takes the place of the DA system, but the lack of regulatory mechanisms may lead to the appearance of LID (Pagano et al., 2017). The same authors developed a meta-analysis from results obtained in PET studies performed in PD patients with and without dyskinesia that showed no differences in SERT binding in the putamen and globus pallidus but a significant reduction in the caudate in dyskinetic compared to non-dyskinetic patients (Pagano et al., 2017).

Again, studies in animal models of LID have provided heterogeneous results. In 6-OHDA lesioned rats rendered dyskinetic by prolonged levodopa administration, lesions of the DRN prevented the expression of dyskinesia (Carta, Carlsson, Kirik, & Bjorklund, 2007; Eskow, Gupta, Alam, Park, & Bishop, 2007). However, in dyskinetic animals, SERT expression has been found to be upregulated (Rylander et al., 2010), not modified (Prinz, Selesnew, Liss, Roeper, & Carlsson, 2013), or decreased (Nevalainen, Af Bjerkén, Lundblad, Gerhardt, & Strömberg, 2011). Interestingly, protein analysis of striatal SERT and DA transporter (DAT) in unilateral sham or 6-OHDA lesioned rats treated with levodopa showed an enhanced SERT:DAT ratio in lesioned rats, which was positively correlated with LID expression (Conti et al., 2016). A comparison between the *in vivo* imaging of neurotransmission, structural markers and immunohistochemistry in MPTP-treated macaques showed that LID severity was linked to increased SERT binding in the ventral anterior cingulate cortex (Beaudoin-Gobert et al., 2018). 5-HT receptor expression in the BG is unevenly modified by levodopa treatment: 5-HT_{2A} and 5-HT_{1B} receptor expression is increased (Riahi, Morissette, Parent, & Di Paolo, 2011; Zhang, Andren, Greengard, & Svenningsson, 2008), while 5-HT_{1A} receptor expression does not change (Riahi et al., 2012). Several studies performed in dyskinetic animals observed no changes or decreases in the number of 5-HT neurons (Inden et al., 2012; Rylander et al., 2010; Stansley & Yamamoto, 2014), no modification in neuron activity and intact 5-HT or DA levels in the DRN (Bishop et al., 2012; Miguelez, Navailles, De Deurwaerdère, & Ugedo, 2016). However, as suggested in Miguelez et al. (2017) it is important to consider that levodopa-derived DA can directly and indirectly modify SERT function, since both levodopa-derived DA and 5-HT compete for this transporter. Indeed, chronic levodopa administration decreases the efficacy of serotonin reuptake inhibitors in behavioral studies in rats (Fidalgo et al., 2015; Miguelez, Berrocoso, Mico, & Ugedo, 2013).

Thus far, only amantadine (a NMDA receptor antagonist) is available for treating LID, which lacks long-term efficiency. Considering the role of the 5-HT in PD together with the lack of efficient treatments for LID, 5-HT-based therapies have been extensively investigated in an attempt to find efficacious anti-dyskinetic drugs. Numerous 5-HT drugs have been demonstrated to improve dyskinesia, although not always without worsening the anti-parkinsonian effect (Bibbiani, Oh, & Chase, 2001; Dupre, Eskow, Negron, & Bishop, 2007; Kleedorfer, Lees, & Stern, 1991). In other cases, LID has been successfully reduced in one animal species but not in others. This is the case for the selective 5-HT_{2A} antagonist EMD-281,014, which reduced dyskinesia in levodopa-treated parkinsonian marmosets without impacting the anti-parkinsonian action of levodopa (Hamadjida et al., 2018), but the same treatment failed to reduce LID in 6-OHDA lesioned rats (Frouni et al., 2019). The main problem is that until now, despite showing anti-dyskinetic properties in preclinical studies, none of these drugs have achieved sufficient good results in clinical trials to be approved for use in patients. For example, sarizotan, a 5-HT_{1A} receptor agonist, improves dyskinesia in macaque (Grégoire et al., 2009) and rat models (Gerlach, Bartoszyk, Riederer, Dean, & van den Buuse, 2011). However, a phase III clinical trial in PD patients failed because sarizotan did not counteract LIDs better than placebo (Goetz et al., 2007). Eltoprazine, a 5-$HT_{1A/B}$-receptor agonist, has proven anti-dyskinetic effects on 6-OHDA lesioned rats treated with levodopa (Lindgren, Andersson, Lagerkvist, Nissbrandt, & Cenci, 2010; Paolone, Brugnoli, Arcuri, Mercatelli, & Morari, 2015) and macaques (Bezard et al., 2013). Eltoprazine also restored synaptic changes and molecular features in the striatum of dyskinetic rats (Ghiglieri et al., 2016). In a double-blind randomized placebo-controlled and dose finding phase I/IIa study conducted in 22 patients, a single dose showed anti-dyskinetic effects (Svenningsson et al., 2015). However, despite of these promising results, no new clinical data have been published since 2015. Similarly, a partial agonist of 5-HT_{1A} receptors, buspirone, has demonstrated strong anti-dyskinetic properties in numerous animal models of LID (Aristieta et al., 2012; Azkona et al., 2014; Dekundy, Lundblad, Danysz, & Cenci, 2007; Eskow et al., 2007; Gerlach, Beck, Riederer, & van den Buuse, 2011) and clinical efficiency in open-label trials performed in humans (Bonifati, Fabrizio, Cipriani, Vanacore, & Meco,

1994; Kleedorfer et al., 1991; Politis et al., 2014) where the reduction on dyskinesia was associated with modulation of striatal DA levels quantified by imaging techniques (Politis et al., 2014).

In summary, the well-accepted role of the 5-HT system in the transformation of levodopa together with the increasing understanding of the adaptive changes that occur during LID development make the 5-HT system a good target for developing new anti-dyskinetic treatments.

6. Conclusions

Although PD is described as a disease of DA neurons, it is well accepted that it is a multisystem disorder. Here, we provide data regarding the alterations of the 5-HT system due to toxicity and adaptive changes during PD and LID (Fig. 1). Numerous evidence indicates the role of the 5-HT system in the motor and non-motor symptoms of PD and LID, suggesting that it could be a good target for therapies for PD and LID. However, the complex mechanisms involved in PD and LID are not yet well determined, and the numerous subtypes of 5-HT receptors expressed in the organism complicate the identification of the proper 5-HT drugs.

Acknowledgments

Supported by: the Spanish government, MINECO, SAF2016-77758-R (AEI/FEDER, UE), the government of the Basque Country T747-13, IT-901/16 and the University of Basque Country PPG 17/51. E.P.-R. has a predoctoral fellowship from the Basque government and S.V. has a predoctoral fellowship from the UPV/EHU.

Conflict of interest

No potential conflicts of interest.

References

Ahlskog, J. E., & Muenter, M. D. (2001). Frequency of levodopa-related dyskinesias and motor fluctuations as estimated from the cumulative literature. *Movement Disorders: Official Journal of the Movement Disorder Society*, *16*(3), 448–458. Retrieved from http://www.ncbi.nlm.nih.gov/pubmed/11391738.

Alarcón-Arís, D., Recasens, A., Galofré, M., Carballo-Carbajal, I., Zacchi, N., Ruiz-Bronchal, E., et al. (2018). Selective α-synuclein knockdown in monoamine neurons by intranasal oligonucleotide delivery: Potential therapy for Parkinson's disease. *Molecular Therapy*, *26*(2), 550–567. https://doi.org/10.1016/j.ymthe.2017.11.015.

Alexander, G. E., DeLong, M. R., & Strick, P. L. (1986). Parallel organization of functionally segregated circuits linking basal ganglia and cortex. *Annual Review of Neuroscience*, *9*(1), 357–381. https://doi.org/10.1146/annurev.ne.09.030186.002041.

Aristieta, A., Azkona, G., Sagarduy, A., Miguelez, C., Ruiz-Ortega, J. Á., Sanchez-Pernaute, R., et al. (2012). The role of the subthalamic nucleus in L-DOPA induced dyskinesia in 6-hydroxydopamine lesioned rats. *PLoS One*, 7(8), e42652. https://doi.org/10.1371/journal.pone.0042652.

Aristieta, A., Morera-Herreras, T., Ruiz-Ortega, J. A., Miguelez, C., Vidaurrazaga, I., Arrue, A., et al. (2014). Modulation of the subthalamic nucleus activity by serotonergic agents and fluoxetine administration. *Psychopharmacology*, *231*(9), 1913–1924. https://doi.org/10.1007/s00213-013-3333-0.

Aristieta, A., Ruiz-Ortega, J. A., Miguelez, C., Morera-Herreras, T., & Ugedo, L. (2016). Chronic L-DOPA administration increases the firing rate but does not reverse enhanced slow frequency oscillatory activity and synchronization in substantia nigra pars reticulata neurons from 6-hydroxydopamine-lesioned rats. *Neurobiology of Disease*, *89*, 88–100. https://doi.org/10.1016/j.nbd.2016.02.003.

Azkona, G., Sagarduy, A., Aristieta, A., Vazquez, N., Zubillaga, V., Ruíz-Ortega, J. A., et al. (2014). Buspirone anti-dyskinetic effect is correlated with temporal normalization of dysregulated striatal DRD1 signalling in l-DOPA-treated rats. *Neuropharmacology*, *79*, 726–737. https://doi.org/10.1016/j.neuropharm.2013.11.024.

Azmitia, E. C., & Nixon, R. (2008). Dystrophic serotonergic axons in neurodegenerative diseases. *Brain Research*, *1217*, 185–194. https://doi.org/10.1016/j.brainres.2008.03.060.

Ballanger, B., Strafella, A. P., van Eimeren, T., Zurowski, M., Rusjan, P. M., Houle, S., et al. (2010). Serotonin 2A receptors and visual hallucinations in parkinson disease. *Archives of Neurology*, *67*(4), 416–421. https://doi.org/10.1001/archneurol.2010.35.

Beaudoin-Gobert, M., Météreau, E., Duperrier, S., Thobois, S., Tremblay, L., & Sgambato, V. (2018). Pathophysiology of levodopa-induced dyskinesia: Insights from multimodal imaging and immunohistochemistry in non-human primates. *NeuroImage*, *183*, 132–141. https://doi.org/10.1016/j.neuroimage.2018.08.016.

Bendor, J. T., Logan, T. P., & Edwards, R. H. (2013). The Function of α-synuclein. *Neuron*, *79*(6), 1044–1066. https://doi.org/10.1016/j.neuron.2013.09.004.

Beucke, J. C., Plotkin, M., Winter, C., Endrass, T., Amthauer, H., Juckel, G., et al. (2011). Midbrain serotonin transporters in de novo and L-DOPA-treated patients with early Parkinson's disease—A [123I]-ADAM SPECT study. *European Journal of Neurology*, *18*(5), 750–755. https://doi.org/10.1111/j.1468-1331.2010.03274.x.

Bezard, E., Tronci, E., Pioli, E. Y., Li, Q., Porras, G., Björklund, A., et al. (2013). Study of the antidyskinetic effect of eltoprazine in animal models of levodopa-induced dyskinesia. *Movement Disorders: Official Journal of the Movement Disorder Society*, *28*(8), 1088–1096. https://doi.org/10.1002/mds.25366.

Bibbiani, F., Oh, J. D., & Chase, T. N. (2001). Serotonin 5-HT1A agonist improves motor complications in rodent and primate parkinsonian models. *Neurology*, *57*(10), 1829–1834. Retrieve from http://www.ncbi.nlm.nih.gov/pubmed/11723272.

Bilen, Ş., Saka, M., Ak, F., & Öztekin, N. (2008). Persistent dystonia induced by fluoxetine. *Internal Medicine Journal*, *38*(8), 672–674. https://doi.org/10.1111/j.1445-5994.2008.01747.x.

Bishop, C., George, J. A., Buchta, W., Goldenberg, A. A., Mohamed, M., Dickinson, S. O., et al. (2012). Serotonin transporter inhibition attenuates l -DOPA-induced dyskinesia without compromising l -DOPA efficacy in hemi-parkinsonian rats. *European Journal of Neuroscience*, *36*(6), 2839–2848. https://doi.org/10.1111/j.1460-9568.2012.08202.x.

Bonifati, V., Fabrizio, E., Cipriani, R., Vanacore, N., & Meco, G. (1994). Buspirone in levodopa-induced dyskinesias. *Clinical Neuropharmacology*, *17*(1), 73–82. Retrieved from http://www.ncbi.nlm.nih.gov/pubmed/8149361.

Braak, H., Rüb, U., Gai, W. P., & Del Tredici, K. (2003). Idiopathic Parkinson's disease: Possible routes by which vulnerable neuronal types may be subject to neuroinvasion by an unknown pathogen. *Journal of Neural Transmission, 110*(5), 517–536. https://doi.org/10.1007/s00702-002-0808-2.

Calon, F., Morissette, M., Rajput, A. H., Hornykiewicz, O., Bédard, P. J., & Di Paolo, T. (2003). Changes of GABA receptors and dopamine turnover in the postmortem brains of parkinsonians with levodopa-induced motor complications. *Movement Disorders, 18*(3), 241–253. https://doi.org/10.1002/mds.10343.

Caretti, V., Stoffers, D., Winogrodzka, A., Isaias, I.-U., Costantino, G., Pezzoli, G., et al. (2008). Loss of thalamic serotonin transporters in early drug-naïve Parkinson's disease patients is associated with tremor: An [(123)I]beta-CIT SPECT study. *Journal of Neural Transmission (Vienna, Austria: 1996), 115*(5), 721–729. https://doi.org/10.1007/s00702-007-0015-2.

Carta, M., Carlsson, T., Kirik, D., & Bjorklund, A. (2007). Dopamine released from 5-HT terminals is the cause of L-DOPA-induced dyskinesia in parkinsonian rats. *Brain, 130*(7), 1819–1833. https://doi.org/10.1093/brain/awm082.

Conti, M. M., Meadows, S. M., Melikhov-Sosin, M., Lindenbach, D., Hallmark, J., Werner, D. F., et al. (2016). Monoamine transporter contributions to l -DOPA effects in hemi-parkinsonian rats. *Neuropharmacology, 110*, 125–134. https://doi.org/10.1016/j.neuropharm.2016.07.025.

Cummings, J., Isaacson, S., Mills, R., Williams, H., Chi-Burris, K., Corbett, A., et al. (2014). Pimavanserin for patients with Parkinson's disease psychosis: A randomised, placebo-controlled phase 3 trial. *Lancet (London, England), 383*(9916), 533–540. https://doi.org/10.1016/S0140-6736(13)62106-6.

Dallé, E., Daniels, W. M. U., & Mabandla, M. V. (2017). Fluvoxamine maleate effects on dopamine signaling in the prefrontal cortex of stressed Parkinsonian rats: Implications for learning and memory. *Brain Research Bulletin, 132*, 75–81. https://doi.org/10.1016/j.brainresbull.2017.05.014.

Dekundy, A., Lundblad, M., Danysz, W., & Cenci, M. A. (2007). Modulation of l-DOPA-induced abnormal involuntary movements by clinically tested compounds: Further validation of the rat dyskinesia model. *Behavioural Brain Research, 179*(1), 76–89. https://doi.org/10.1016/j.bbr.2007.01.013.

Doder, M., Rabiner, E. A., Turjanski, N., Lees, A. J., Brooks, D. J., & 11C-WAY 100635 PET study. (2003). Tremor in Parkinson's disease and serotonergic dysfunction: An 11C-WAY 100635 PET study. *Neurology, 60*(4), 601–605. Retrieved from http://www.ncbi.nlm.nih.gov/pubmed/12601099.

Dorsey, E. R., & Bloem, B. R. (2018). The Parkinson pandemic—A call to action. *JAMA Neurology, 75*(1), 9. https://doi.org/10.1001/jamaneurol.2017.3299.

Dupre, K. B., Eskow, K. L., Negron, G., & Bishop, C. (2007). The differential effects of 5-HT1A receptor stimulation on dopamine receptor-mediated abnormal involuntary movements and rotations in the primed hemiparkinsonian rat. *Brain Research, 1158*, 135–143. https://doi.org/10.1016/j.brainres.2007.05.005.

Eskow, K. L., Gupta, V., Alam, S., Park, J. Y., & Bishop, C. (2007). The partial 5-HT1A agonist buspirone reduces the expression and development of l-DOPA-induced dyskinesia in rats and improves l-DOPA efficacy. *Pharmacology Biochemistry and Behavior, 87*(3), 306–314. https://doi.org/10.1016/j.pbb.2007.05.002.

Fidalgo, C., Ko, W. K. D., Tronci, E., Li, Q., Stancampiano, R., Chuan, Q., et al. (2015). Effect of serotonin transporter blockade on L-DOPA-induced dyskinesia in animal models of Parkinson's disease. *Neuroscience, 298*, 389–396. https://doi.org/10.1016/j.neuroscience.2015.04.027.

Fox, S. H. (2013). Non-dopaminergic treatments for motor control in Parkinson's disease. *Drugs, 73*(13), 1405–1415. https://doi.org/10.1007/s40265-013-0105-4.

Frouni, I., Kwan, C., Bédard, D., Belliveau, S., Bourgeois-Cayer, É., Gaudette, F., et al. (2019). Effect of the selective 5-HT2A receptor antagonist EMD-281,014 on l-DOPA-induced abnormal involuntary movements in the 6-OHDA-lesioned rat. *Experimental Brain Research*, *237*, 29–36. https://doi.org/10.1007/s00221-018-5390-4.

Gagnon, D., Eid, L., Coudé, D., Whissel, C., Di Paolo, T., Parent, A., et al. (2018). Evidence for sprouting of dopamine and serotonin axons in the pallidum of Parkinsonian monkeys. *Frontiers in Neuroanatomy*, *12*, 38. https://doi.org/10.3389/fnana.2018.00038.

Gagnon, D., Gregoire, L., Di Paolo, T., & Parent, M. (2016). Serotonin hyperinnervation of the striatum with high synaptic incidence in parkinsonian monkeys. *Brain Structure & Function*, *221*(7), 3675–3691. https://doi.org/10.1007/s00429-015-1125-5.

Gerlach, M., Bartoszyk, G. D., Riederer, P., Dean, O., & van den Buuse, M. (2011). Role of dopamine D3 and serotonin 5-HT1A receptors in l-DOPA-induced dyskinesias and effects of sarizotan in the 6-hydroxydopamine-lesioned rat model of Parkinson's disease. *Journal of Neural Transmission*, *118*(12), 1733–1742. https://doi.org/10.1007/s00702-010-0571-8.

Gerlach, M., Beck, J., Riederer, P., & van den Buuse, M. (2011). Flibanserin attenuates l-DOPA sensitized contraversive circling in the unilaterally 6-hydroxydopamine-lesioned rat model of Parkinson's disease. *Journal of Neural Transmission*, *118*(12), 1727–1732. https://doi.org/10.1007/s00702-010-0570-9.

Ghiglieri, V., Mineo, D., Vannelli, A., Cacace, F., Mancini, M., Pendolino, V., et al. (2016). Modulation of serotonergic transmission by eltoprazine in L-DOPA-induced dyskinesia: Behavioral, molecular, and synaptic mechanisms. *Neurobiology of Disease*, *86*, 140–153. https://doi.org/10.1016/j.nbd.2015.11.022.

Goetz, C. G., Damier, P., Hicking, C., Laska, E., Müller, T., Olanow, C. W., et al. (2007). Sarizotan as a treatment for dyskinesias in Parkinson's disease: A double-blind placebo-controlled trial. *Movement Disorders*, *22*(2), 179–186. https://doi.org/10.1002/mds. 21226.

Grégoire, L., Samadi, P., Graham, J., Bédard, P. J., Bartoszyk, G. D., & Di Paolo, T. (2009). Low doses of sarizotan reduce dyskinesias and maintain antiparkinsonian efficacy of l-Dopa in parkinsonian monkeys. *Parkinsonism & Related Disorders*, *15*(6), 445–452. https://doi.org/10.1016/J.PARKRELDIS.2008.11.001.

Halliday, G. M., Blumbergs, P. C., Cotton, R. G., Blessing, W. W., & Geffen, L. B. (1990). Loss of brainstem serotonin- and substance P-containing neurons in Parkinson's disease. *Brain Research*, *510*(1), 104–107. Retrieved from http://www.ncbi.nlm.nih.gov/pubmed/1691042.

Hamadjida, A., Nuara, S. G., Bédard, D., Gaudette, F., Beaudry, F., Gourdon, J. C., et al. (2018). The highly selective 5-HT 2A antagonist EMD-281,014 reduces dyskinesia and psychosis in the l -DOPA-treated parkinsonian marmoset. *Neuropharmacology*, *139*, 61–67. https://doi.org/10.1016/j.neuropharm.2018.06.038.

Hernandez, D. G., Reed, X., & Singleton, A. B. (2016). Genetics in Parkinson disease: Mendelian versus non-Mendelian inheritance. *Journal of Neurochemistry*, *139*, 59–74. https://doi.org/10.1111/jnc.13593.

Huang, X., Yang, J., Yang, S., Cao, S., Qin, D., Zhou, Y., et al. (2017). Role of tandospirone, a 5-HT1A receptor partial agonist, in the treatment of central nervous system disorders and the underlying mechanisms. *Oncotarget*, *8*(60), 102705–102720. https://doi.org/10.18632/oncotarget.22170.

Huot, P., Johnston, T. H., Koprich, J. B., Fox, S. H., & Brotchie, J. M. (2013). The pharmacology of L-DOPA-induced dyskinesia in Parkinson's disease. *Pharmacological Reviews*, *65*(1), 171–222. https://doi.org/10.1124/pr.111.005678.

Huot, P., Johnston, T. H., Visanji, N. P., Darr, T., Pires, D., Hazrati, L.-N., et al. (2012). Increased levels of 5-HT1A receptor binding in ventral visual pathways in Parkinson's disease. *Movement Disorders: Official Journal of the Movement Disorder Society*, *27*(6), 735–742. https://doi.org/10.1002/mds.24964.

Iadanza, M. G., Jackson, M. P., Hewitt, E. W., Ranson, N. A., & Radford, S. E. (2018). A new era for understanding amyloid structures and disease. *Nature Reviews Molecular Cell Biology*, *19*(12), 755–773. https://doi.org/10.1038/s41580-018-0060-8.

Inden, M., Abe, M., Minamino, H., Takata, K., Yoshimoto, K., Tooyama, I., et al. (2012). Effect of selective serotonin reuptake inhibitors via 5-HT1A receptors on L-DOPA-induced rotational behavior in a hemiparkinsonian rat model. *Journal of Pharmacological Sciences*, *119*(1), 10–19. Retrieved from http://www.ncbi.nlm.nih.gov/pubmed/22510520.

Jinsmaa, Y., Cooney, A., Sullivan, P., Sharabi, Y., & Goldstein, D. S. (2015). The serotonin aldehyde, 5-HIAL, oligomerizes alpha-synuclein. *Neuroscience Letters*, *590*, 134–137. https://doi.org/10.1016/j.neulet.2015.01.064.

Joling, M., van den Heuvel, O. A., Berendse, H. W., Booij, J., & Vriend, C. (2018). Serotonin transporter binding and anxiety symptoms in Parkinson's disease. *Journal of Neurology, Neurosurgery, and Psychiatry*, *89*(1), 89–94. https://doi.org/10.1136/jnnp-2017-316193.

Kalia, L. V., & Lang, A. E. (2015). Parkinson's disease. *Lancet (London, England)*, *386*(9996), 896–912. https://doi.org/10.1016/S0140-6736(14)61393-3.

Kish, S. J., Tong, J., Hornykiewicz, O., Rajput, A., Chang, L.-J., Guttman, M., et al. (2008). Preferential loss of serotonin markers in caudate versus putamen in Parkinson's disease. *Brain*, *131*(Pt. 1), 120–131. https://doi.org/10.1093/brain/awm239.

Kiss, J., Shannak, K., & Hornykiewicz, O. (1988). Uneven pattern of dopamine loss in the striatum of patients with idiopatic Parkinson´s disease. *New England Journal of Medicine*, *318*, 876–880. Retrieved from https://doi.org/10.1056/NEJM198804073181402.

Kleedorfer, B., Lees, A. J., & Stern, G. M. (1991). Buspirone in the treatment of levodopa induced dyskinesias. *Journal of Neurology, Neurosurgery, and Psychiatry*, *54*(4), 376–377. Retrieved from http://www.ncbi.nlm.nih.gov/pubmed/2056334.

Kohl, Z., Ben Abdallah, N., Vogelgsang, J., Tischer, L., Deusser, J., Amato, D., et al. (2016). Severely impaired hippocampal neurogenesis associates with an early serotonergic deficit in a BAC α-synuclein transgenic rat model of Parkinson's disease. *Neurobiology of Disease*, *85*, 206–217. https://doi.org/10.1016/j.nbd.2015.10.021.

Kotagal, V., Spino, C., Bohnen, N. I., Koeppe, R., & Albin, R. L. (2018). Serotonin, β-amyloid, and cognition in Parkinson disease. *Annals of Neurology*, *83*(5), 994–1002. https://doi.org/10.1002/ana.25236.

Kovacs, G. G., Klöppel, S., Fischer, I., Dorner, S., Lindeck-Pozza, E., Birner, P., et al. (2003). Nucleus-specific alteration of raphe neurons in human neurodegenerative disorders. *Neuroreport*, *14*(1), 73–76. https://doi.org/10.1097/01.wnr.0000050301.92401.0c.

Lee, L. J., & Lee, L. J. H. (2012). Neonatal fluoxetine exposure alters motor performances of adolescent rats. *Developmental Neurobiology*, *72*(8), 1122–1132. https://doi.org/10.1002/dneu.20942.

Leo, R. J. (1996). Movement disorders associated with the serotonin selective reuptake inhibitors. *The Journal of Clinical Psychiatry*, *57*(10), 449–454. Retrieved from http://www.ncbi.nlm.nih.gov/pubmed/8909330.

Lim, J., Bang, Y., Choi, J. H., Han, A., Kwon, M. S., Liu, K. H., et al. (2018). LRRK2 G2019S Induces Anxiety/Depression-like behavior before the onset of motor dysfuncion with 5-HT1A receptor upregulation in mice. *Journal of Neuroscience*, *38*(7), 1611–1621. https://doi.org/10.1523/JNEUROSCI.4051-15.2017.

Lindgren, H. S., Andersson, D. R., Lagerkvist, S., Nissbrandt, H., & Cenci, M. A. (2010). l-DOPA-induced dopamine efflux in the striatum and the substantia nigra in a rat model of Parkinson's disease: Temporal and quantitative relationship to the expression of dyskinesia. *Journal of Neurochemistry*, *112*(6), 1465–1476. https://doi.org/10.1111/j.1471-4159.2009.06556.x.

Loane, C., Wu, K., Bain, P., Brooks, D. J., Piccini, P., & Politis, M. (2013). Serotonergic loss in motor circuitries correlates with severity of action-postural tremor in PD. *Neurology*, *80*(20), 1850–1855. https://doi.org/10.1212/WNL.0b013e318292a31d.

Maillet, A., Krack, P., Lhommée, E., Météreau, E., Klinger, H., Favre, E., et al. (2016). The prominent role of serotonergic degeneration in apathy, anxiety and depression in *de novo* Parkinson's disease. *Brain*, *139*(9), 2486–2502. https://doi.org/10.1093/brain/aww162.
Meltzer, H. Y., Mills, R., Revell, S., Williams, H., Johnson, A., Bahr, D., et al. (2010). Pimavanserin, a serotonin2a receptor inverse agonist, for the treatment of Parkinson's disease psychosis. *Neuropsychopharmacology*, *35*(4), 881–892. https://doi.org/10.1038/npp.2009.176.
Miguelez, C., Benazzouz, A., Ugedo, L., & De Deurwaerdère, P. (2017). Impairment of serotonergic transmission by the antiparkinsonian drug L-DOPA: Mechanisms and clinical implications. *Frontiers in Cellular Neuroscience*, *11*, 274. https://doi.org/10.3389/fncel.2017.00274.
Miguelez, C., Berrocoso, E., Mico, J. A., & Ugedo, L. (2013). l-DOPA modifies the antidepressant-like effects of reboxetine and fluoxetine in rats. *Neuropharmacology*, *67*, 349–358. https://doi.org/10.1016/j.neuropharm.2012.11.016.
Miguelez, C., Morera-Herreras, T., Torrecilla, M., Ruiz-Ortega, J. A., & Ugedo, L. (2014). Interaction between the 5-HT system and the basal ganglia: Functional implication and therapeutic perspective in Parkinson's disease. *Frontiers in Neural Circuits*, *8*, 21. https://doi.org/10.3389/fncir.2014.00021.
Miguelez, C., Navailles, S., De Deurwaerdère, P., & Ugedo, L. (2016). The acute and long-term L-DOPA effects are independent from changes in the activity of dorsal raphe serotonergic neurons in 6-OHDA lesioned rats. *British Journal of Pharmacology*, *173*(13), 2135–2146. https://doi.org/10.1111/bph.13447.
Navailles, S., Bioulac, B., Gross, C., & De Deurwaerdère, P. (2011). Chronic L-DOPA therapy alters central serotonergic function and L-DOPA-induced dopamine release in a region-dependent manner in a rat model of Parkinson's disease. *Neurobiology of Disease*, *41*(2), 585–590. https://doi.org/10.1016/j.nbd.2010.11.007.
Nevalainen, N., Af Bjerkén, S., Lundblad, M., Gerhardt, G. A., & Strömberg, I. (2011). Dopamine release from serotonergic nerve fibers is reduced in L-DOPA-induced dyskinesia. *Journal of Neurochemistry*, *118*(1), 12–23. https://doi.org/10.1111/j.1471-4159.2011.07292.x.
Obeso, J. A., Rodríguez-Oroz, M. C., Benitez-Temino, B., Blesa, F. J., Guridi, J., Marin, C., et al. (2008). Functional organization of the basal ganglia: Therapeutic implications for Parkinson's disease. *Movement Disorders*, *23*(S3), S548–S559. https://doi.org/10.1002/mds.22062.
Pagano, G., Yousaf, T., & Politis, M. (2017). PET molecular imaging research of levodopa-induced dyskinesias in Parkinson's disease. *Current Neurology and Neuroscience Reports*, *17*(11), 90. https://doi.org/10.1007/s11910-017-0794-2.
Paolone, G., Brugnoli, A., Arcuri, L., Mercatelli, D., & Morari, M. (2015). Eltoprazine prevents levodopa-induced dyskinesias by reducing striatal glutamate and direct pathway activity. *Movement Disorders*, *30*(13), 1728–1738. https://doi.org/10.1002/mds.26326.
Pavese, N., Metta, V., Bose, S. K., Chaudhuri, K. R., & Brooks, D. J. (2010). Fatigue in Parkinson's disease is linked to striatal and limbic serotonergic dysfunction. *Brain*, *133*(11), 3434–3443. https://doi.org/10.1093/brain/awq268.
Petrou, M., Dwamena, B. A., Foerster, B. R., MacEachern, M. P., Bohnen, N. I., Müller, M. L., et al. (2015). Amyloid deposition in Parkinson's disease and cognitive impairment: A systematic review. *Movement Disorders*, *30*(7), 928–935. https://doi.org/10.1002/mds.26191.
Politis, M., Loane, C., Wu, K., Brooks, D. J., & Piccini, P. (2011). Serotonergic mediated body mass index changes in Parkinson's disease. *Neurobiology of Disease*, *43*(3), 609–615. https://doi.org/10.1016/j.nbd.2011.05.009.
Politis, M., Wu, K., Loane, C., Brooks, D. J., Kiferle, L., Turkheimer, F. E., et al. (2014). Serotonergic mechanisms responsible for levodopa-induced dyskinesias in Parkinson's disease patients. *The Journal of Clinical Investigation*, *124*(3), 1340–1349. https://doi.org/10.1172/JCI71640.

Politis, M., Wu, K., Loane, C., Kiferle, L., Molloy, S., Brooks, D. J., et al. (2010). Staging of serotonergic dysfunction in Parkinson's disease: An in vivo 11C-DASB PET study. *Neurobiology of Disease*, *40*(1), 216–221. https://doi.org/10.1016/j.nbd.2010.05.028.

Politis, M., Wu, K., Loane, C., Quinn, N. P., Brooks, D. J., Oertel, W. H., et al. (2012). Serotonin neuron loss and nonmotor symptoms continue in Parkinson's patients treated with dopamine grafts. *Science Translational Medicine*, *4*(128), 128ra41. https://doi.org/10.1126/scitranslmed.3003391.

Prinz, A., Selesnew, L.-M., Liss, B., Roeper, J., & Carlsson, T. (2013). Increased excitability in serotonin neurons in the dorsal raphe nucleus in the 6-OHDA mouse model of Parkinson's disease. *Experimental Neurology*, *248*, 236–245. https://doi.org/10.1016/j.expneurol.2013.06.015.

Qamhawi, Z., Towey, D., Shah, B., Pagano, G., Seibyl, J., Marek, K., et al. (2015). Clinical correlates of raphe serotonergic dysfunction in early Parkinson's disease. *Brain*, *138*(10), 2964–2973. https://doi.org/10.1093/brain/awv215.

Riahi, G., Morissette, M., Parent, M., & Di Paolo, T. (2011). Brain 5-HT2A receptors in MPTP monkeys and levodopa-induced dyskinesias. *European Journal of Neuroscience*, *33*(10), 1823–1831. https://doi.org/10.1111/j.1460-9568.2011.07675.x.

Riahi, G., Morissette, M., Lévesque, D., Rouillard, C., Samadi, P., Parent, M., et al. (2012). Effect of chronic l-DOPA treatment on 5-HT(1A) receptors in parkinsonian monkey brain. *Neurochem Int.*, *61*(7), 1160–1171. https://doi.org/10.1016/j.neuint.2012.08.009.

Rylander, D., Parent, M., O'Sullivan, S. S., Dovero, S., Lees, A. J., Bezard, E., et al. (2010). Maladaptive plasticity of serotonin axon terminals in levodopa-induced dyskinesia. *Annals of Neurology*, *68*(5), 619–628. https://doi.org/10.1002/ana.22097.

Sagarduy, A., Llorente, J., Miguelez, C., Morera-Herreras, T., Ruiz-Ortega, J. A., & Ugedo, L. (2016). Buspirone requires the intact nigrostriatal pathway to reduce the activity of the subthalamic nucleus via 5-HT 1A receptors. *Experimental Neurology*, *277*, 35–45. https://doi.org/10.1016/j.expneurol.2015.12.005.

Shannak, K., Rajput, A., Rozdilsky, B., Kish, S., Gilbert, J., & Hornykiewicz, O. (1994). Noradrenaline, dopamine and serotonin levels and metabolism in the human hypothalamus: observations in Parkinson's disease and normal subjects. *Brain Research*, *639*(1), 33–41. Retrieved from http://www.ncbi.nlm.nih.gov/pubmed/8180836.

Spillantini, M. G., Schmidt, M. L., Lee, V. M.-Y., Trojanowski, J. Q., Jakes, R., & Goedert, M. (1997). α-Synuclein in lewy bodies. *Nature*, *388*(6645), 839–840. https://doi.org/10.1038/42166.

Stansley, B. J., & Yamamoto, B. K. (2014). Chronic L-dopa decreases serotonin neurons in a subregion of the dorsal raphe nucleus. *Journal of Pharmacology and Experimental Therapeutics*, *351*(2), 440–447. https://doi.org/10.1124/jpet.114.218966.

Stansley, B. J., & Yamamoto, B. K. (2015a). Behavioral impairments and serotonin reductions in rats after chronic L-dopa. *Psychopharmacology*, *232*(17), 3203–3213. https://doi.org/10.1007/s00213-015-3980-4.

Stansley, B., & Yamamoto, B. (2015b). L-Dopa and brain serotonin system dysfunction. *Toxics*, *3*(1), 75–88. https://doi.org/10.3390/toxics3010075.

Sulzer, D. (2010). Clues to how alpha-synuclein damages neurons in Parkinson's disease. *Movement Disorders*, *25*(S1), S27–S31. https://doi.org/10.1002/mds.22639.

Svenningsson, P., Rosenblad, C., Af Edholm Arvidsson, K., Wictorin, K., Keywood, C., Shankar, B., et al. (2015). Eltoprazine counteracts l-DOPA-induced dyskinesias in Parkinson's disease: A dose-finding study. *Brain*, *138*(4), 963–973. https://doi.org/10.1093/brain/awu409.

Takahashi, M., Tabu, H., Ozaki, A., Hamano, T., Takeshima, T., & REBORN study group. (2019). Antidepressants for depression, apathy, and gait instability in Parkinson's disease: A multicenter randomized study. *Internal Medicine (Tokyo, Japan)*, *58*, 361–368. https://doi.org/10.2169/internalmedicine.1359-18.

Tatara, A., Shimizu, S., Shin, N., Sato, M., Sugiuchi, T., Imaki, J., et al. (2012). Modulation of antipsychotic-induced extrapyramidal side effects by medications for mood disorders. *Progress in Neuro-Psychopharmacology and Biological Psychiatry*, *38*(2), 252–259. https://doi.org/10.1016/j.pnpbp.2012.04.008.

Tong, Q., Zhang, L., Yuan, Y., Jiang, S., Zhang, R., Xu, Q., et al. (2015). Reduced plasma serotonin and 5-hydroxyindoleacetic acid levels in Parkinson's disease are associated with nonmotor symptoms. *Parkinsonism & Related Disorders*, *21*(8), 882–887. https://doi.org/10.1016/j.parkreldis.2015.05.016.

Wan, O. W., Shin, E., Mattsson, B., Caudal, D., Svenningsson, P., & Björklund, A. (2016). α-Synuclein induced toxicity in brain stem serotonin neurons mediated by an AAV vector driven by the tryptophan hydroxylase promoter. *Scientific Reports*, *6*(1), 26285. https://doi.org/10.1038/srep26285.

Wersinger, C., Rusnak, M., & Sidhu, A. (2006). Modulation of the trafficking of the human serotonin transporter by human alpha-synuclein. *European Journal of Neuroscience*, *24*(1), 55–64. https://doi.org/10.1111/j.1460-9568.2006.04900.x.

Wersinger, C., & Sidhu, A. (2009). Partial regulation of serotonin transporter function by gamma-synuclein. *Neuroscience Letters*, *453*(3), 157–161. https://doi.org/10.1016/j.neulet.2009.02.033.

Wilson, H., Giordano, B., Turkheimer, F. E., Chaudhuri, K. R., & Politis, M. (2018). Serotonergic dysregulation is linked to sleep problems in Parkinson's disease. *NeuroImage. Clinical*, *18*, 630–637. https://doi.org/10.1016/j.nicl.2018.03.001.

Ye, Z., Altena, E., Nombela, C., Housden, C. R., Maxwell, H., Rittman, T., et al. (2014). Selective serotonin reuptake inhibition modulates response inhibition in Parkinson's disease. *Brain*, *137*(4), 1145–1155. https://doi.org/10.1093/brain/awu032.

Zhang, X., Andren, P. E., Greengard, P., & Svenningsson, P. (2008). Evidence for a role of the 5-HT1B receptor and its adaptor protein, p11, in L-DOPA treatment of an animal model of Parkinsonism. *Proceedings of the National Academy of Sciences*, *105*(6), 2163–2168. https://doi.org/10.1073/pnas.0711839105.

Zuo, L.-J., Yu, S.-Y., Hu, Y., Wang, F., Piao, Y.-S., Lian, T.-H., et al. (2016). Serotonergic dysfunctions and abnormal iron metabolism: Relevant to mental fatigue of Parkinson disease. *Scientific Reports*, *6*(1), 19. https://doi.org/10.1038/s41598-016-0018-z.

CHAPTER SEVEN

New therapeutic approaches to target alpha-synuclein in Parkinson's disease: The role of immunotherapy☆

T. Fernández-Valle[a,b,*], I. Gabilondo[a,b], J.C. Gómez-Esteban[a,b]

[a]Movement Disorders and Autonomic Unit, Neurology Department, Cruces University Hospital, Barakaldo, Spain
[b]Neurodegenerative Unit, Biocruces Health Research Institute, Barakaldo, Spain
*Corresponding author: e-mail address: tamara.fernandezvalle@osakidetza.eus

Contents

Abstract

Parkinson's disease (PD) is the second most prevalent neurodegenerative disorder. It is characterized by a slow and progressive loss of dopaminergic neurons.

Its neuropathological hallmark is the accumulation of aggregated form of α-synuclein (α-syn) protein in intracellular inclusions known as Lewy bodies. This aggregated α-syn is believed to be central to the pathogenesis of PD. Emerging evidence suggests that aggregated forms of α-syn self-amplificates and propagates spreading from cell-to-cell in a "prion-like" fashion.

Genetics and environmental factors are known causes for the pathogenesis of PD.

In last years, inflammation in the pathophysiology of PD is gaining more importance. This neuroinflammation seems to contribute to the progressive degeneration of dopaminergic neurons.

The currently available therapies for PD fail to modify the disease progression and neurodegeneration. The connection between α-syn and PD makes α-syn the major

☆This article contains no third-party or previously published material.

International Review of Neurobiology, Volume 146
ISSN 0074-7742
https://doi.org/10.1016/bs.irn.2019.06.014

therapeutic target. We summarize the possible therapeutic strategies to target α-syn according to the steps in the molecular pathogenesis. The contribution of neuroinflammation to the progression of the disease and the "prion-like" hypothesis which enables targeting the extracellular phase of transmission of α-syn, make immunotherapy probably the most promising therapeutic approach for PD.

1. Introduction

Parkinson's disease (PD) is a neurodegenerative disorder characterized by motor and non- motor (psychiatric, cognitive, sleep and autonomic) features.

This disease is the second most prevalent neurodegenerative disorder after Alzheimer's disease, affecting 7–10 million persons worldwide.

PD is characterized by a slow and progressive loss of dopaminergic neurons in the *substantia nigra*, as well as other types of neurons, in the central, peripheral and enteric nervous system.(Poewe, Seppi, & Tanner, 2017).

Genetics and environmental factors are some of the known causes for the pathogenesis of PD.

The neuropathological hallmark of PD and some other neurodegenerative disorders is the accumulation of aggregated form of α-synuclein protein (α-syn) in intracellular inclusions known as Lewy bodies and Lewy neurites. These diseases are known as synucleinopathies and also include Multiple System Atrophy, Lewy Bodies Dementia and Pure Autonomic Failure (Spillantini, Schmidt, Lee, et al., 1997).

The normal function of α-syn is unknown but it is believed to take part in neurons synaptic transmission, mostly in the recycling of synaptic vesicles (Burré, Sharma, Südhof, et al., 2018).The protein is also abundant in red blood cells although its functions are not yet well understood.

Normal α-syn is bound to de membrane of synaptic vesicles and it is present in the cytosol as a soluble and unfolded monomer and it seems that also as a tetramer. Under physiological conditions there is a balance between the monomeric forms and the tetramers of this protein (Dettmer, Newman, Soldner, et al., 2015).

Under certain conditions, α-syn misfolds and forms insoluble aggregates that are the pathological hallmarks of synucleinopathies. These pathological forms are organized in protofibrils, often phosphorylated and nitrated, although these last biochemical changes could be an epiphenomenon (Dettmer et al., 2015).

Missense mutations in the gene encoding α-syn (SNCA) and duplication or triplication of SNCA gene are associated with autosomal dominant forms of PD (Brundin, Dave, Kordower, et al., 2017; Nalls, Pankratz, Lill, et al., 2014; Weihofen, Liu, Arndt, et al., 2019). Moreover, SNCA has been identified as one of the strongest risk loci related with sporadic PD (Nalls et al., 2014). Sporadic forms of PD are also associated with other genes not encoding α-syn but that also result in α-syn accumulation suggesting that those mutations might exacerbate the role of other factors such as oxidation, nitration and decreased lisosomal functions. Normal aging is also associated with increased levels of cytoplasmic soluble α-syn. Furthermore, a strong correlation between clinical manifestations of the disease and the presence and the severity of α-syn aggregations in the brain and peripheral nerves has been described (Braak, Del Tredici, Rüb, et al., 2003). As α-syn levels are increased, a possible therapeutic intervention could be decreasing the level of α-syn.

Emerging evidence in laboratory models suggests that aggregated forms of α-syn self-amplificates and propagates spreading from cell-to-cell over interconnected neural networks in a stereotypical and topographical pattern from peripheral tissues to the brain in a "prion-like" fashion (Visanji, Brooks, Hazrati, et al., 2013). This has been observed in embryonic nigral transplants in PD patients after a decade of the initial grafting (Kordower, Chu, Hauser, et al., 2008). In the same way, it has been reported that intracerebral injection of aggregated α-syn accelerates the onset of neurological symptoms in transgenic mice expressing human α-syn (Luk, Kehm, & Zhang, 2012).

As we have pointed before, it is well recognized that Lewy pathology does not develop at the same time in all parts of the nervous system at that it stars in peripheral tissues. Evidence suggests that olfactory bulb and dorsal motor nucleus of the vagus are two areas where Lewy bodies appear before affecting the brain (Braak & Del Tredici, 2017). Recent publications even imply that the first affected structures would be outside the CNS, in the small nerve fibers of the autonomic nervous system (Coon, Cutsforth-Gregory, & Benarroch, 2018), which regulate functions such as intestinal transit or the variability of blood pressure or heart rate. This fact is important for an early therapeutic intervention because these areas are likely to cause the prodromal symptoms and signs.

Inflammation in the pathophysiology of PD is gaining more importance in last years. The relationship between α-syn misfolding and neuroinflammation is complex. This neuroinflammation seems to contribute to the progressive degeneration of dopaminergic neurons although the exact mechanism in not fully understood.

The first evidence for the involvement of the immune system came from the observation of activated microglia surrounding degenerating dopaminergic neurons in the SNpc. Recent research focused in microglia which is the mayor scavenger for extracellular α-syn aggregated. An excessive microglial activation has been observed with the resulting increase in inflammatory response mediated by proinflammatory cytokines including tumor necrosis factor alpha (TNF- α), interleukin 1-β (IL 1-β), interleukin-6 (IL-6) and interferon γ (INF- γ). The microglial activation induced changes in both the efficacy to take up extracellular α -syn and their ability to degrade it (Zhang, Xia, Wan, et al., 2018). Also, neuron-derived exosomes were found to transfer α -syn toxic forms between neural and non-neural cell (microglia) contributing to spreading (Chistiakov & Chistiakov, 2017).

From observation of activated microglia of post-mortem brain of PD patients has been described that accumulation of α-syn activates the microglia and promotes neuroinflammation with the participation of innate and adaptative immunity. Numerous studies have reported elevated levels of proinflammatory mediators in the cerebrospinal fluid and brain tissue (striatum, substantia nigra) (Duke, Moran, Pearce, & Graeber, 2007; Taylor, Main, & Crack, 2013). It is also suggested that the activated neuroimmune response can promote post-translational modifications in mechanism such as oxidation and nitration of α-syn which triggers misfolding of the protein (Gao, Kotzbauer, Uryu, et al., 2008). In recent years, although not consistently, it has been implicated that the use of non-steroidal anti-inflammatory drugs could decrease the incidence of Parkinson's disease in populations that use these drugs (Gagne & Power, 2010; Samii, Etminan, Wiens, & Jafari, 2009; Wahner, Bronstein, Bordelon, & Ritz, 2007).

Recent advances in the field of genomic, have identified genes encoding immune system molecules that are related with other pathologies and that affect PD risk (Sivakumaran, Agakov, Theodoratou, et al., 2011). This is the case of multifunctional leucine-rich repeat kinase 2 (LRRK2) gen. Some of its variants are recognized as risk factors for PD and some variants are also independently linked to inflammatory bowel disease (Crohn's disease and Ulcerative Colitis). Experimental data suggest that LRRK2 is an important modulator in the immune system, pathophysiologically linking PD and inflammatory bowel disease (IBD),(Peter, Dubinsky, Bressman, et al., 2018).

In a couple of studies made analyzing retrospectively databases of patients with IBD, it has been reported a higher incidence of PD amount IBD compared with age, sex matched controls. Furthermore, one of these studies showed that patients with IBD who were prescribed anti-TNF had a lower risk of developing PD than those patients with IBD who were not taking anti-TNF. This fact could mean that a reduction of systemic inflammation could be the reason of the observed association.

The available therapies for PD target only dopamine-related deficiencies with little effect on non-dopaminergic features and with some debilitating adverse effects. These treatments fail to modify the disease progression and neurodegeneration.

The connection of α-syn to disease pathology and progression makes it one of the most compelling targets for the development of disease modifying therapies. The aim of this article is to summarize the possible therapeutic strategies that are currently being proposed (Table 1).

Table 1 Resume of ongoing clinical trials against α-syn in Parkinson's disease.

Intervention	Sponsor	Mechanism	Type of study	Status	Results
Affitope PD01A Affitope PD03A	AFFiRis	Vaccine	Phase 1/2	Completed (AFF008; AFF011)	Pending
PRX 002 (R0704615)	Roche/ Prothena	Monoclonal antibody	Phase 2	Active, not recruiting	
BIIB054	Biogen	Monoclonal antibody	Phase 2	Recruiting	
NPT200-11	Neuropore Therapies/ UCB	α-Syn aggregation modulator	Phase 1b	Planned	
ANLE 138b	MODAG	α-Syn aggregation modulator	Phase 1	Planned	
NPT088	Proclana Biosciences	α-Syn aggregation modulator	Preclinic	Planned	
Ambroxol	Exar Corporation	Gcase enhancer chaperone	Phase 2	Recruiting	
GZ/SAR4027671	Sanofi Genzime	Gcase enhancer	Phase 2	Recruiting	
Nilotinib	Novartis	Inhibits protein c-abl	Phase 2	Active, not recruiting	
BAN0805 (ABBV-0805)	BioArtic/ AbbVie	Antibody	Phase 1	Planned	

2. Reducing α-synuclein synthesis

Since it has been described that increased levels of α-syn are related with PD, decreasing production might be a logical therapeutic approach. Reducing the cytoplasmic levels of α-syn also decreases the risk that the protein adopts an abnormal conformation and aggregates.

Decreasing production of α-syn at a time prior to its aggregation might prevent its aggregation and rescue the function and viability of neuronal populations that are vulnerable in PD.

RNA molecules targeting α-syn via RNA interference have been tested in neuron-like cell cultures and rodent models in vivo. Direct infusion of small interfering RNA (siRNA) reduced hippocampal and cortical α-syn expression in mice. There is also evidence that systemic exosomal siRNA decreased aggregation of α-syn and reduced synthesis of α-syn via microRNAs (miRNAs), siRNA, antisense oligonucleotides (ASO) or other transcription inhibitor making them potential therapeutic options (Lewis, Melrose, Bumcrot, et al., 2008).

Unilateral siRNA infusions against α-syn in monkeys reduced levels of α-syn 40–50% compared with the untreated side (McCormack, Mak, Henderson, et al., 2010).

The delivery of short hairpin α-syn RNA (shRNA) inhibited 35% of the expression of endogenous α-syn in the rats (Zharikov, Cannon, Tapias, et al., 2015). There was found no toxicity in rats.

Since normal α-syn is considered to have important functions in the CNS, we must be cautious with therapies which reduce dramatically the levels of α-syn. In experiments which have achieved extensive loss of α-syn, neurotoxicity has been observed. It has been reported that the knockdown of endogenous α-syn (reductions >90%) in the substancia nigra of rats and non-human primates produces the degeneration of the nigrostriatal system (Collier, Redmond, Steece-Collier, et al., 2016; Gorbatyuk, Li, Nash, et al., 2010; Kannan & Manfredsson, 2012). Furthermore, if the therapies are not specifically targeted to the brain, they could have similar effects in peripheral tissues (e.g., peripheral nerves, red blood cells) whose consequences are not yet well understood.

A safe degree of reduction will have to be established prior to clinical trial.

Another possible target could be to find a chemical compound designed to reduce the transcription of the SNCA gene. β2-Adrenoreceptor (β2AR) has been identified as a regulator of the SNCA and β2AR ligands modulate SNCA transcription. Mechanistically, β2AR ligands were shown to modulate transcription through histone H3 lysine27 acetylation of its promoters and enhancers. In two epidemiological studies in Norway has been seen that salbutamol (a brain-penetrant β2AR agonist use to treat asthma) reduced risk of developing PD. Conversely a β2AR antagonist (propranolol) was associated with an increased risk ***(Mittal, Bjornevik, Im, et al., 2017). However, another recent article has not found this association (Searles Nielsen, Gross, Camacho-Soto, et al., 2018).

3. Preventing α-synuclein aggregation

Inhibiting α-syn aggregation seems a great approach since this way normal α-syn can be preserved and the toxicity that follows misfolding can be prevented. It has been challenging, however, to develop effective strategies to suppress the aggregation of α-syn due to its mechanism of aggregation which is extremely complex and dependent on environmental factors.

Heat shock proteins (HSP) are molecular chaperones that prevent protein aggregation. They act to stabilize partially folded protein intermediates, in an ATP-independent manner, to maintain cellular proteostasis under stress conditions. A variety of them, particularly HSP70 and HSP104, have been effective in reducing α-syn aggregation in vitro and in vivo (Gorenberg & Chandra, 2017). For example, in dopaminergic neurons of substantia nigra there is a relatively low basal level of this chaperones and makes them more susceptible to protein aggregation. To date, there are no clinical trials (Brundin, Dave, Kordower, et al., 2017).

Another approach is the use on intrabodies which are small antibody fragments which targets antigens intracellularly. They can bind α-syn monomers and prevent them from oligomerizing. Anti-α-syn nanobodies fused to a proteasome-targeting proline, aspartate or glutamate, serine, and threonine motif can modulate monomeric concentrations of α-syn. They have many advantages of conventional antibodies, including high specificity and affinity for target epitopes. Different antibodies have shown promise in vitro. It appears that intrabodies may be neuroprotective interfering with the aggregation prone region (NAC). Two examples are, nanobody VH14*PEST and nanobody NbSyn87. Both of them completely eliminated aggregated α-syn, restored striatal dopamine, and improved motor function. These data suggest that intrabodies may be effective at lowering levels of soluble α-syn and thereby preventing α -syn aggregation. A challenge for the clinical application is going to be how to achieve high levels in the CNS for prolonged periods. Unlike immunotherapies, intrabodies will require direct CNS delivery using viral vectors which are been characterized (Brundin et al., 2017).

The oligomer modulator anle138b was tested to be neuroprotective (Wagner, Ryazanov, Leonov, et al., 2013). Furthermore, many new aggregation inhibitors such a as curcumin have shown neuroprotective properties (Wang, Boddapati, Emadi, & Sierks, 2010).

NPT200-11 and NPT088 are two candidates that are currently in clinical testing phase. NPT200-11 has been reported to prevent the formation of toxic oligomeric aggregates and improve behavioral, neuropathological and biochemical outcome measures in α-syn overexpressing transgenic mouse model. NPT088 is a fusion protein combining human immunoglobulin backbone as well as a General Amyloid Interaction Motif (or GAIM) and has been reported to bind α-syn aggregates, reduce the accumulation of proteinase K-resistant protein (Zhang et al., 2018).

Recent evidence suggests that therapeutic strategies directed at the prevention of α-syn-mediated microglial activation, either by blocking direct interaction with surface receptors on microglia or interrupting the microglial-mediated inflammation cascades, may prevent downstream pathogenesis (Joers, Tansey, Mulas, & Carta, 2017).

4. Promoting degradation of intracellular α-synuclein aggregates

Autophagy-lysosomal pathway plays an important role in the degradation of intracellular α-syn. Modulating the lysosomal enzyme β-glucocerebrosidase (GBA enzyme) could be an interesting therapeutic approach. GBA is a lysosomal enzyme responsible for the metabolism of glucocerebroside (GC). GBA deficiency causes accumulation of GC and also α-syn in the lysosomes. Glucocerebrosidase (GBA1) gene is the one which encodes GCase enzyme. Mutations in this gene (GBA1) constitutes the major genetic risk factor for PD. The proportion of PD patients that carry GBA1 mutations is estimated to be 7–10%, and that these mutations confer a 20- to 30-fold increased risk of developing PD with penetrance estimated up to 30% at 80 years of age. These mutations have been reported to reduce GBA enzyme activity in PD brain tissue and cerebrospinal fluid.

The strong genetic data and the mechanistic link to α-syn have made GBA an attractive target for PD drug development. There is currently a phase 2 trial (MOVES-PD) testing the efficacy of the compound GZ/SAR402671 in GBA1-mutation carriers (Jankovic, Goodman, Safirstein, et al., 2018). Other two ongoing phase 2 trials are testing safety, tolerability and efficacy of Ambroxol in PD. Ambroxol, an FDA-approved mucolytic, acts as a chaperone and have shown to improve lysosomal function in cells with GBA mutations in vitro and increases GBA enzyme activity in non-human primates in vivo (Silveira, MacKinley, Coleman, et al., 2019).

Nilotinib is approved for chronic myelogenous leukemia. It inhibits protein target c-abl, which is an Abelson murine leukemia viral oncogene found in cells. Several studies have showed that c-abl activity is enhanced in brain tissue of PD patients and that this increased c-abl activity leads to a downstream increase in phosphorylation and aggregation of α-syn. Increased c-abl activity also reduced function of Parkin, a key protein involved in mitochondrial biogenesis (Lonskaya, Hebron, & Desforges, 2014). In animal models of PD has been proved that pharmacological and genetic manipulations to reduce c-abl activity have neuroprotective effects. C-abl inhibitors have limitation in treating CNS disorders because of pharmacokinetics and small therapeutic window for chronic usage. A recent small, open-label study (lacking a placebo control) in patients suffering from PD and dementia or DLB reported that nilotinib was safe and tolerable at doses much lower than those used against cancer (Pagan, Hebron, & Valadez, 2016). More studies are needed.

Rapamycin is an antibiotic that has therapeutic potential of autophagy-enhancing agent in Parkinson's disease. The first studies in 1-methyl-4-phenyl-1,2,3,6-tetrahydropyridine (MPTP) mouse model of PD seem to demonstrate an increase in GBA enzyme activity and decreased neurodegeneration in immunohistochemical studies.

5. Increasing degradation of extracellular α-synuclein. Immunotherapies against α-synuclein

The discovery of inflammation in the pathophysiology of PD and the role of microglia and immune system make immunotherapies probably the most promising therapeutic approach for PD.

Antibodies are unable to enter cells, so they target extracellular α-syn.

Active immunization stimulates the immune system to produce antibodies against target proteins while for passive immunization the antibodies are directly administered conferring a temporal protection. Both active and passive immunization have presented neuroprotective effects in animal models. Both have been studied in α-syn transgenic mice and have demonstrate that anti-α-syn antibodies with high relative affinity to the C-terminus protein region decrease intracellular accumulation of α-syn in cell bodies and synapses, protect against synaptic loss and gliosis, and ameliorate motor and cognitive behavior deficits (Jankovic et al., 2018).

Immunotherapy has also an anti-inflammatory effect. It induces microglia activation and reduces the production of pro-inflammatory cytokines.

The company Affiris has recently announced the result of a Phase I clinical trial with an active immunotherapy vaccine called AFFITOPE PD01A and PD03A in patients with Multiple System Atrophy. AFFITOPE® PD01A and PD03A are synthetically produced vaccines containing α-syn-mimicking peptide active. They were well tolerated in early MSA patients. The majority of adverse events were mild and local reactions. PD01A induced a clear immune response versus the peptide itself and α-Syn epitope. (Affiris Company). There is another pilot Phase 1 randomized, placebo-controlled study which assessed the tolerability and safety of repeated subcutaneous (under the skin) injections of two doses of Affitope PD03A in people with early Parkinson's. Results of the trial showed that both doses were locally and systemically well-tolerated. There was no drug-related serious adverse event. The treatment also showed a dose-dependent immune response against itself and cross-reactivity against the α-syn-targeted epitope over time, showing antibody reactivation 9 months after the first dose, with the booster immunization.

Two passive immunotherapy approaches have progressed to clinical trials: Prothena (PRX002) and Biogen (BIIB-054).

Prothena is testing PRX002 which is a humanized IgG1 monoclonal antibody designed to target the C-terminus of neurotoxic (aggregated) forms of α-syn. Passive immunotherapy was tested in a phase 1 trial with reduction of unbound serum α-syn levels up to 96%. The phase 2 trial is now in process (PASADENA-ROCHE) with the latest update that single and multiple doses of PRX002 were safe and well tolerated and resulted in robust binding of peripheral α-syn and dose-dependent increases of PRX002 in CSF.

Biogen is using the BIB-054 antibody which is a recombinant human IgG1 antibody directed against N terminal epitope on α-syn and is highly selective for aggregated forms of α-syn. The results are similar to PRX002 antibody. The phase 1 demonstrated good tolerability, favorable pharmacokinetic profile and achieved levels in CSF. The phase 2 trial is now ongoing in individuals with early PD (SPARK-BIOGEN).

Because of the differences in their selectivity for oligomeric and fibrillary α-syn aggregates relative to α-syn monomer, clinical data from BIB054 and PRX002 will provide valuable insights into the optimal product profile for an α-syn therapeutic antibody. (Weihofen, Liu, Arndt, et al., 2019).

Additional clinical immunotherapy programs are on the horizon. That is the case of BAN0805 which is another clinical-phase antibody that targets oligomeric form of α-syn that are thought to be pathogenic (BioArtic Company) (Fagerqvist, Lindström, Nordström, et al., 2013).

Finally, in in vitro and in animal models has been seen that Granulocyte-macrophage colony stimulating factor (GM-CSF) acts broadly to differentiate dendritic cells (DCs) and affect immune transformation from effector to regulatory immune responses. DCs skew such immune responses by increasing regulatory T-cell (Treg) numbers and activities that serve to attenuate proinflammatory responses and augment neuroprotection (Schutt, Gendelman, & Mosley, 2018).

In a recent phase 1 clinical trial, treatment of PD patients with human recombinant GM-CSF (sargramostim) increased Treg numbers and Treg-mediated suppression of CD3/CD28-induced proliferation compared to placebo-treated PD patients. Moreover, sargramostim improved motor deficits and initiation of motor activity as determined by improvement in UPDRS III scores with paralleled improved magneto encephalography-recorded cortical motor activities. These data underscore the potential for immune transformation by GM-CSF to diminish neurodegeneration in PD (Gendelman, Zhang, Santamaria, et al., 2017).

Immunotherapies have still some limitation. They can trigger other targets and provoke non-specific inflammatory response. Furthermore, repetitive administration is needed, and the penetration of antibodies can be limited.

6. Challenges facing clinical trials with α-synuclein therapies

α-Syn is a complex protein with extreme conformational diversity, which molecular specie is the best target remains unknown. Moreover, α-syn may adopt different conformation depending on environment conditions and from patient to patient.

Another challenge is the lack of biomarkers that would allow an early diagnosis and treatment. The ideal approach would be to treat PD in the prodromal phase.

Monitoring the response is also difficult cause is a slow degenerative disorder and clinical trial's duration is not long enough.

7. Conclusion

The role of α-syn in the disease pathology and progression makes it one of the most relevant targets for the development of disease modifying therapies in PD and other synucleinopathies. We have explained the different possible strategies to target α-syn according to the steps in the molecular pathogenesis (Fig. 1). Neuroinflammation contribute to the progressive degeneration of dopaminergic neurons by releasing factors that could enhance oxidative stress, protein misfolding and aggregation. The "prion-like" hypothesis in which α-syn is responsible for initiating and spreading cell-to-cell the pathological process in PD enable a possible therapeutic strategy targeting the extracellular phase of transmission of α-syn. For all these reasons, among all the possible strategies, immunotherapy is probably the most promising therapeutic approach for PD.

However, although α-syn as therapeutic target is a promising approach, there are challenge that must be face up. The complexity of α-syn and the

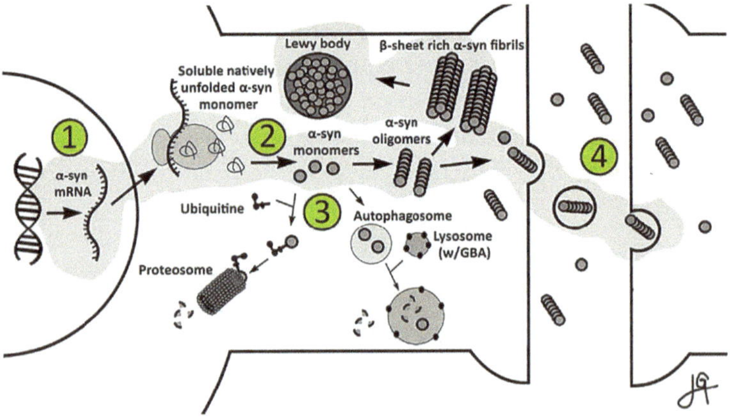

Fig. 1 Different steps to target α-syn in the molecular pathogenesis: 1. Reducing α-synuclein synthesis, 2. preventing α-synuclein aggregation, 3. promoting degradation of intracellular α-synuclein aggregates, 4. increasing degradation of extracellular α-synuclein.

incomplete understanding of the precise role of α-syn in the disease development are facts that limit the development of therapeutic strategies. Further research will be needed to face these challenges in order to develop more purified therapeutic strategies.

References

Braak, H., & Del Tredici, K. (2017). Neuropathological staging of brain pathology in sporadic Parkinson's disease: Separating the wheat from the chaff. *Journal of Parkinson's disease*, 7(s1), S71–S85.

Braak, H., Del Tredici, K., Rüb, U., et al. (2003). Staging of brain pathology related to sporadic Parkinson's disease. *Neurobiology of Aging*, *24*(2), 197–211.

Brundin, P., Dave, K. D., Kordower, J., et al. (2017). Therapeutic approaches to target alpha-synuclein pathology. *Experimental Neurology*, *298*, 225–235.

Burré, J., Sharma, M., Südhof, T. C., et al. (2018). Cell biology and pathophysiology of α-synuclein. *Cold Spring Harbor Perspectives in Medicine*, *8*(3), a024091–29.

Chistiakov, D. A., & Chistiakov, A. A. (2017). α-Synuclein-carrying extracellular vesicles in Parkinson's disease: Deadly transmitters. *Acta Neurologica Belgica*, *117*(1), 43–51.

Collier, T., Redmond, D., Steece-Collier, K., et al. (2016). Is alpha-synuclein loss-of-function a contributor to parkinsonian pathology? evidence from non-human primates. *Frontiers in Neuroscience*, *10*, 12.

Coon, E. A., Cutsforth-Gregory, J. K., & Benarroch, E. E. (2018). Neuropathology of autonomic dysfunction in synucleinopathies. *Movement Disorders*, *33*, 349–358.

Dettmer, U., Newman, A., Soldner, F., et al. (2015). Parkinson-causing α-synuclein missense mutations shift native tetramers to monomers as a mechanism for disease initiation. *Nature Communications*, *6*, 7314.

Duke, D. C., Moran, L. B., Pearce, R. K., & Graeber, M. B. (2007). The medial and lateral substantia nigra in Parkinson's disease: mRNA profiles associated with higher brain tissue vulnerability. *Neurogenetics*, *8*, 83–94.

Fagerqvist, T., Lindström, V., Nordström, E., et al. (2013). Monoclonal antibodies selective for α-synuclein oligomers/protofibrils recognize brain pathology in Lewy body disorders and α-synuclein transgenic mice with the disease-causing A30P mutation. *Journal of Neurochemistry*, *126*(1), 131–144.

Gagne, J. J., & Power, M. C. (2010). Anti-inflammatory drugs and risk of Parkinson disease: A meta-analysis. *Neurology*, *74*(12), 995–1002.

Gao, H. M., Kotzbauer, P. T., Uryu, K., et al. (2008). Neuroinflammation and oxidation/nitration of alpha-synuclein linked to dopaminergic neurodegeneration. *The Journal of Neuroscience*, *28*(30), 7687–7698.

Gendelman, H. E., Zhang, Y., Santamaria, P., et al. (2017). Evaluation of the safety and immunomodulatory effects of sargramostim in a randomized, double-blind phase 1 clinical Parkinson's disease trial. *NPJ Parkinson's disease*, *3*, 10.

Gorbatyuk, O., Li, S., Nash, K., et al. (2010). In vivo RNAi-mediated alpha-synuclein silencing induces nigrostriatal degeneration. *Molecular Therapy*, *18*, 1450–1457.

Gorenberg, E. L., & Chandra, S. S. (2017). The role of co-chaperones in synaptic proteostasis and neurodegenerative disease. *Frontiers in Neuroscience*, *11*, 248.

Jankovic, J., Goodman, I., Safirstein, B., et al. (2018). Safety and tolerability of multiple ascending doses of PRX002/RG7935, an anti-α-synuclein monoclonal antibody, in patients with Parkinson disease: A randomized clinical trial. *JAMA Neurology*, *75*(10), 1206–1214.

Joers, V., Tansey, M. G., Mulas, G., & Carta, A. R. (2017). Microglial phenotypes in Parkinson's disease and animal models of the disease. *Progress in Neurobiology*, *155*, 57–75.

Kannan, N., & Manfredsson, F. (2012). Loss of functional alpha-synuclein: A toxic event in Parkinson's disease? *Journal of Parkinson's disease, 2*, 249–267.

Kordower, J. H., Chu, Y., Hauser, R. A., et al. (2008). Lewy body-like pathology in long-term embryonic nigral transplants in Parkinson's disease. *Nature Medicine, 14*(5), 504–506.

Lewis, J., Melrose, H., Bumcrot, D., et al. (2008). In vivo silencing of alpha-synuclein using naked siRNA. *Molecular Neurodegeneration, 3*, 19.

Lonskaya, I., Hebron, M. L., & Desforges, N. M. (2014). Nilotinib-induced autophagic changes increase endogenous parkin level and ubiquitination, leading to amyloid clearance. *Journal of Molecular Medicine (Berlin, Germany), 92*(4), 373–386.

Luk, K. C., Kehm, V. M., & Zhang, B. (2012). Intracerebral inoculation of pathological α-synuclein initiates a rapidly progressive neurodegenerative α-synucleinopathy in mice. *The Journal of Experimental Medicine, 209*(5), 975–986.

McCormack, A., Mak, S., Henderson, J., et al. (2010). Alpha-synuclein suppression by targeted small interfering RNA in the primate substantia nigra. *PLoS One, 5*(8), e12122.

Mittal, S., Bjornevik, K., Im, D., et al. (2017). B2-Adrenoreceptor is a regulator of the α-synuclein gene driving risk of Parkinson's disease. *Science, 357*, 891–898.

Nalls, M. A., Pankratz, N., Lill, C. M., et al. (2014). Large-scale meta-analysis of genome-wide association data identifies six new risk loci for Parkinson's disease. *Nature Genetics, 46*(9), 989–993.

Pagan, F., Hebron, M., & Valadez, E. H. (2016). Nilotinib effects in Parkinson's disease and dementia with Lewy bodies. *Journal of Parkinson's disease, 6*(3), 503–517.

Peter, I., Dubinsky, M., Bressman, S., et al. (2018). Anti-tumor necrosis factor therapy and incidence of Parkinson disease among patients with inflammatory bowel disease. *JAMA Neurology, 75*(8), 939–946.

Samii, A., Etminan, M., Wiens, M. O., & Jafari, S. (2009). NSAID use and the risk of Parkinson's disease: systematic review and meta-analysis of observational studies. *Drugs & Aging, 26*(9), 769–779.

Schutt, C. R., Gendelman, H. E., & Mosley, R. L. (2018). Tolerogenic bone marrow-derived dendritic cells induce neuroprotective regulatory T cells in a model of Parkinson's disease. *Molecular Neurodegeneration, 13*(1), 26.

Searles Nielsen, S., Gross, A., Camacho-Soto, A., et al. (2018). β2-adrenoreceptor medications and risk of Parkinson disease. *Annals of Neurology, 84*(5), 683–693.

Silveira, C., MacKinley, J., Coleman, K., et al. (2019). Ambroxol as a novel disease-modifying treatment for Parkinson's disease dementia: protocol for a single-centre, randomized, double-blind, placebo-controlled trial. *BMC Neurology, 19*, 20.

Sivakumaran, S., Agakov, F., Theodoratou, E., et al. (2011). Abundant pleiotropy in human complex diseases and traits. *American Journal of Human Genetics, 89*(5), 607–618.

Spillantini, M. G., Schmidt, M. L., Lee, V. M., et al. (1997). Alpha synuclein in Lewy bodies. *Nature, 388*, 839–840.

Taylor, J. M., Main, B. S., & Crack, P. J. (2013). Neuroinflammation and oxidative stress: Co-conspirators in the pathology of Parkinson's disease. *Neurochemistry International, 62*, 803–819.

Visanji, N. P., Brooks, P. L., Hazrati, L.-N., et al. (2013). The prion hypothesis in Parkinson's disease: Braak to the future. *Acta Neuropathologica Communications, 1*, 2.

Poewe, W., Seppi, K., & Tanner, C. M. (2017). Parkinson disease. *Nature Reviews. Disease Primers, 3* 17013.

Wagner, J., Ryazanov, S., Leonov, A., et al. (2013). Anle138b: a novel oligomer modulator for disease-modifying therapy of neurodegenerative diseases such as prion and Parkinson's disease. *Acta Neuropathologica, 125*(6), 795–813.

Wahner, A. D., Bronstein, J. M., Bordelon, Y. M., & Ritz, B. (2007). Nonsteroidal anti-inflammatory drugs may protect against Parkinson disease. *Neurology, 69*(19), 1836–1842.

Wang, M. S., Boddapati, S., Emadi, S., & Sierks, M. R. (2010). Curcumin reduces alpha-synuclein induced cytotoxicity in Parkinson's disease cell model. *BMC Neuroscience*, *11*, 57.

Weihofen, A., Liu, Y., Arndt, J., et al. (2019). Development of an aggregate-selective, human-derived α-synuclein antibody NIIB054 that ameliorates disease phenotypes in Parkinson's disease models. *Neurobiology of Disease*, *124*, 276–288.

Zhang, G., Xia, Y., Wan, F., et al. (2018). New perspectives on roles of alpha-synuclein in Parkinson's disease. *Frontiers in Aging Neuroscience*, *10*, 370.

Zharikov, A., Cannon, J., Tapias, V., et al. (2015). shRNA targeting alpha-synuclein prevents neurodegeneration in a Parkinson's disease model. *The Journal of Clinical Investigation*, *125*, 2721–2735.

FSC
www.fsc.org
MIX
Papier | Fördert gute Waldnutzung
FSC® C083411

Kolibri 360 GmbH
Ferdinand-Jühlke-Straße 7
99095 Erfurt, Deutschland
produktsicherheit@kolibri360.de